St. Mary's High School

THE LUCENT LIBRARY OF SCIENCE AND TECHNOLOGY

Space Stations

by James Barter

LUCENT BOOKS®

THOMSON
GALE

San Diego • Detroit • New York • San Francisco • Cleveland • New Haven, Conn. • Waterville, Maine • London • Munich

On cover: A U.S. space shuttle docks with the International Space Station in orbit above the earth.

© 2004 by Lucent Books. Lucent Books is an imprint of The Gale Group, Inc., a division of Thomson Learning, Inc.

Lucent Books® and Thomson Learning™ are trademarks used herein under license.

For more information, contact
Lucent Books
27500 Drake Rd.
Farmington Hills, MI 48331-3535
Or you can visit our Internet site at http://www.gale.com

ALL RIGHTS RESERVED.
No part of this work covered by the copyright hereon may be reproduced or used in any form or by any means—graphic, electronic, or mechanical, including photocopying, recording, taping, Web distribution, or information storage retrieval systems—without the written permission of the publisher.

LIBRARY OF CONGRESS CATALOGING-IN-PUBLICATION DATA

Barter, James, 1946–
   Space stations / by James Barter.
      p. cm. — (The Lucent library of science and technology)
   Includes bibliographical references and index.
     ISBN 1-59018-106-9
     1. Space stations—Juvenile literature. [1. Space stations. 2. Astronautics.] I. Title.
II. Series.
    TL797.15.B37 2004
    629.44'2—dc22
                                                                                         2003026301

Printed in the United States of America

Table of Contents

Foreword	4
Introduction	7
The Road to Space Stations	
Chapter 1	15
Modest Beginnings: Salyut and Skylab	
Chapter 2	30
A Quantum Leap in Technology	
Chapter 3	48
Living in Outer Space	
Chapter 4	65
Space Medicine	
Chapter 5	77
Research and Experiments	
Epilogue	94
Have Space Stations Met Expectations?	
Notes	98
For Further Reading	101
Works Consulted	103
Index	107
Picture Credits	112
About the Author	112

Foreword

"The world has changed far more in the past 100 years than in any other century in history. The reason is not political or economic, but technological—technologies that flowed directly from advances in basic science."

— Stephen Hawking, "A Brief History of Relativity," *Time,* 2000

The twentieth-century scientific and technological revolution that British physicist Stephen Hawking describes in the above quote has transformed virtually every aspect of human life at an unprecedented pace. Inventions unimaginable a century ago have not only become commonplace but are now considered necessities of daily life. As science historian James Burke writes, "We live surrounded by objects and systems that we take for granted, but which profoundly affect the way we behave, think, work, play, and in general conduct our lives."

For example, in just one hundred years, transportation systems have dramatically changed. In 1900 the first gasoline-powered motorcar had just been introduced, and only 144 miles of U.S. roads were hard-surfaced. Horse-drawn trolleys still filled the streets of American cities. The airplane had yet to be invented. Today 217 million vehicles speed along 4 million miles of U.S. roads. Humans have flown to the moon and commercial aircraft are capable of transporting passengers across the Atlantic Ocean in less than three hours.

The transformation of communications has been just as dramatic. In 1900 most Americans lived and worked on farms without electricity or mail delivery. Few people had ever heard a radio or spoken on a telephone. A hundred years later, 98 percent of American homes have telephones and televisions and more than 50 percent have personal computers. Some families even have more than one television and computer, and cell phones are now commonplace, even among the young. Data beamed from communication satellites routinely predict global weather conditions and fiber-optic cable, e-mail, and the Internet have made worldwide telecommunication instantaneous.

Perhaps the most striking measure of scientific and technological change can be seen in medicine and public health. At the beginning of the twentieth century, the average American life span was forty-seven years. By the end of the century the average life span was approaching eighty years, thanks to advances in medicine including the development of vaccines and antibiotics, the discovery of powerful diagnostic tools such as X rays, the life-saving technology of cardiac and neonatal care, and improvements in nutrition and the control of infectious disease.

Rapid change is likely to continue throughout the twenty-first century as science reveals more about physical and biological processes such as global warming, viral replication, and electrical conductivity, and as people apply that new knowledge to personal decisions and government policy. Already, for example, an international treaty calls for immediate reductions in industrial and automobile emissions in response to studies that show a potentially dangerous rise in global temperatures is caused by human activity. Taking an active role in determining the direction of future changes depends on education; people must understand the possible uses of scientific research and the effects of the technology that surrounds them.

The Lucent Books Library of Science and Technology profiles key innovations and discoveries that have transformed the modern world. Each title strives to make a complex scientific discovery, technology, or phenomenon understandable and relevant to the reader. Because scientific discovery is rarely straightforward, each title explains the dead ends, fortunate accidents, and basic scientific methods by which the research into the subject proceeded. And every book examines the practical applications of an invention, branch of science, or scientific principle in industry, public health, and personal life, as well as potential future uses and effects based on ongoing research. Fully documented quotations, annotated bibliographies that include both print and electronic sources, glossaries, indexes, and technical illustrations are among the supplemental features designed to point researchers to further exploration of the subject.

Introduction

The Road to Space Stations

Space stations have been a mainstay of human space exploration since the early 1970s. Of the many types of hardware that have been blasted into space—whether orbiting satellites, reusable transport vehicles such as the shuttle, or unmanned robotic probes making one-way journeys to distant planets and beyond—none possess the complexity, technological sophistication, or size of a space station. These qualities make space stations the workhorses of space exploration. They are highly valued as medical laboratories for learning about the effects of weightlessness on humans, platforms for astronomical studies of distant stars and galaxies, and observation posts for viewing and analyzing natural and human-caused phenomena on Earth.

The technological complexity of modern space stations took decades to develop. Today's International Space Station (ISS), the latest and most complex laboratory in space, is the beneficiary of research efforts dating back to the early 1950s, when the first plans were drawn up to blast a simple satellite no larger than a large beach ball into Earth's orbit. The road to space stations began as a dream that quickly captured the imagination of scientists. These researchers' efforts resulted in technological innovations that made possible ever larger and more sophisticated space stations.

Since the late nineteenth century visionaries have dreamed of space stations in orbit around Earth. This artist's conception is from 1957.

The First Dreams

Even in the late nineteenth and early twentieth centuries, space stations were viewed as solutions to many of Earth's problems. Scientists, physicians, science-fiction writers, politicians, industrialists, and some of the general public dreamed about space stations in orbit around Earth. For some, space stations promised to be places to perform cutting-edge scientific research in a weightless environment. Physicians dreamed of performing medical experiments on astronauts, and other scientists hoped to create crystals, semiconductors, and pharmaceuticals with far fewer of the imperfections caused by Earth's gravitational field. For others, space stations promised to serve as staging points for travel to distant planets or even as space colonies that might one day relieve the already apparent overcrowding and pollution on Earth.

The concept of a staffed outpost in Earth's orbit served as grist for science fiction from 1869, when American writer Edward Everett Hale published a story titled "The Brick Moon" in *Atlantic Monthly* magazine. In the article, Hale's manned outpost was intended to function as a navigational aid for ships plying Earth's oceans. More than fifty years later, in 1923, Romanian science-fiction writer Hermann Oberth became the first person to use *space station* as a term for an installation that would serve as the jumping-off place for human journeys to the Moon and to Mars. Just five years later, Herman Noordung, an Austrian scientist, published the first space station blueprint. His design consisted of a doughnut-shaped structure that comprised crew living quarters, a power-generating station attached to one end of the central hub, and an astronomical observation station.

In modern times, the first person to seriously consider the creation of space stations was the German rocket scientist Wernher von Braun. In 1952 von Braun published his concept of a space station in *Collier's* magazine. He envisioned a facility shaped like a wheel that would have a diameter of 250 feet and would orbit more than 1,000 miles above Earth. Space historian Randy Liebermann explains that the *Collier's* article was part of a broad but carefully crafted vision on von Braun's part:

> After 25 years of continuous and directed thinking and endless hours of experimentation, von Braun, the world's leading rocket engineer, had the chance to come out of his sequestered military environment and through a national magazine inform the general public of his detailed blueprint for realizing manned space travel.[1]

Science Energizes the Dream

Von Braun and other scientists took up the challenge of making the dream of using space stations

Wernher von Braun

Wernher von Braun was one of the most important rocket developers and champions of space exploration between the 1930s and the 1970s. Born in Germany in 1912, he became enamored with the possibilities of space exploration as a young boy by reading the science fiction of Jules Verne and H.G. Wells. As a college student, von Braun mastered calculus and trigonometry so he could understand the physics of rocketry. As a means of furthering his interest to build rockets, in 1932 he went to work for the German army to develop missiles. While engaged in this work, von Braun received his PhD in aerospace engineering in 1934.

During World War II von Braun led what has been called the rocket team, which developed the V-2 ballistic missile for the Germans. The brainchild of von Braun's rocket team, the V-2 was an early version of rockets used in space exploration programs in the United States and the Soviet Union. A liquid propellant missile forty-six feet tall and weighing twenty-seven thousand pounds, the V-2 flew at speeds in excess of thirty-five hundred miles per hour and delivered a twenty-two-hundred-pound warhead to a target five hundred miles away.

At the end of the war in 1945, von Braun and five hundred of his top rocket scientists surrendered to the U.S. Army. For fifteen years after World War II, von Braun worked with the army in the development of ballistic missiles; in 1960 his rocket development center transferred from the army to the newly established NASA. Von Braun soon became director of NASA's Marshall Space Flight Center and the chief architect of the Saturn launch vehicle used to propel Americans to the Moon.

Toward the end of his life, von Braun was one of the most prominent spokesmen of space exploration in the United States. In 1970 NASA leadership asked von Braun to move to Washington, D.C., to head up the strategic planning effort for the agency. On June 16, 1977, while still actively working on aerospace projects, he died in his home in Alexandria, Virginia.

German-born Wernher von Braun was one of the foremost rocket scientists of the twentieth century.

The Road to Space Stations

for exploration a reality. But before anything so sophisticated as a space station could have any hope of succeeding, smaller, simpler efforts would be necessary to prove that launch vehicles would function properly, the needed altitudes could be reached, and humans could survive the stresses of launch and return to Earth.

In 1955, as a first step, American president Dwight Eisenhower announced plans to build and launch a 3.5 pound satellite into orbit. This satellite would circle Earth once, taking photographs as it did so. Unbeknownst to President Eisenhower, Premier Nikita Khrushchev of the Soviet Union had secretly issued orders for Soviet engineers to work on a similar project. The Soviets achieved success first, and on October 4, 1957, launched a satellite named *Sputnik*, the Russian world meaning "satellite." The USSR's accomplishment stunned the United States. *Sputnik* weighed an incredible 183 pounds, raising fears among Americans that they had fallen far behind the Soviets in technological and scientific prowess.

The next step was to determine if an animal could survive the launch and live in space. On November 3, 1957, the Soviets launched *Sputnik 2*, which was large enough to accommodate a dog named Laika. The dog survived in space for several days, making the mission an unqualified success and paving the road for the first human in orbit.

The string of Soviet space successes continued when, in 1961, cosmonaut Yury Gagarin blasted into orbit, circling the earth once in a trip lasting 106 minutes. Gagarin's historic test flight proved that Soviet engineers were able to build spacecraft capable of carrying a human into orbit. The success also confirmed that more sophisticated vehicles such as space stations were feasible.

The Space Race

The Americans realized, as did the Soviets, that before a space station that could house astronauts for long periods in orbit could be undertaken, more experimentation was needed. Highest on their list of problems to solve was producing a rocket powerful enough to thrust far heavier payloads into orbit. Also of serious concern were understanding the effect of weightlessness on humans; perfecting pressurized suits needed for survival in the vacuum of outer space; establishing reliable systems for communication between ground crews and astronauts; providing reliable supplies of food, water, and oxygen; and testing systems for a safe return to Earth.

Desperate to overshadow the Soviet successes, the United States chose to solve these problems on a project more daring and complicated than simply building an orbital outpost. On May 25, 1961, newly elected president John F. Kennedy proposed a national objective of a manned lunar landing with these challenging words: "I believe that this nation should commit itself to achieving the goal, before this decade is out, of landing a man on the moon and returning him safely to the Earth."[2]

On July 20, 1969, as hundreds of millions of people around the world watched on television sets, American astronaut Neil A. Armstrong descended from his lunar module, named *Eagle,* and set foot on the Moon. Following twenty-one hours on the lunar surface gathering forty-six pounds of lunar rocks and planting an American flag, Armstrong and two compatriots safely returned to Earth.

From Rockets and Satellites to Space Stations

Once humans had circled Earth and walked on the Moon, neither the United States nor the Soviet Union were much inclined toward exploring farther from Earth. According to Dr. Gary Lofgren, lunar curator and planetary geoscientist at the National Aeronautics and Space Administration's Johnson Space Center, "We found, however, no new elements or any materials significantly different from the earth. This should not be surprising because our solar system was all made from the same stuff. The differences between the planets reflect primarily their distances from the sun and the stability of various materials."[3]

Disappointment over the results of lunar exploration, however, did not dampen interest in space stations among American and Soviet planners. For example, a special team of Soviet engineers busied themselves developing a space station primarily for military purposes. In a highly ambitious and supersecret project, the Soviets envisioned an orbiting outpost equipped with powerful spy cameras, radar, and even self-defense guns. It would also include supply vehicles and multiple reentry capsules.

A military space station was canceled in favor of a scientific project with wider and more practical applications. On February 9, 1970, the Soviet government officially endorsed the space station called Salyut, a Russian word meaning "greetings." As promised, at the beginning of 1970 Salyut, the world's first space station, was readied for launch in 1971.

Meanwhile, American officials backed the idea of a space station with similarly vague goals in mind. Secretary of Defense Robert McNamara publicly stated the need for a space station for defense in the late 1960s, noting:

> The objective will be to test operations in space using equipment and personnel that can also meet some military needs. We propose a manned orbiting laboratory, not for a precise clearly defined military mission, but because we believe it useful to develop some technologies that could prove essential for manned military operations in space.[4]

To move the program forward as quickly as possible, engineers searched for something that could be

Following the success of Neil Armstrong's 1969 lunar landing, engineers began to develop a space station that would serve as both a research facility and a base for military operations.

adapted to serve as the first space station shell. One recommendation made by many engineers was to convert one of the stages of a Saturn rocket for this purpose. According to Dr. Andrew Dunar and Dr. Stephen Waring, engineers at the Marshall Space Flight Center in Huntsville, Alabama,

> devised schemes for use of a spent-rocket stage as a manned orbiting laboratory that helped form foundations for *Skylab* [America's first space station]. The Research Projects Laboratory conducted studies for science-oriented projects on board *Skylab* including High Energy Astronomy Observatories (HEAO), the Large Space Telescope, and the Apollo Telescope Mount (ATM).[5]

For a combination of reasons that included military, scientific, and economic motives, both the United States and the Soviet Union committed themselves to deployment of space stations.

Chapter 1

Modest Beginnings: Salyut and Skylab

The Soviets won the race to place the first space station in orbit. On April 19, 1971, they launched Salyut 1. Two years later, on May 14, 1973, America responded by launching Skylab. Both space stations shared the modest objectives of studying the effects of long-duration spaceflight on the human body, photographing Earth and the rest of the solar system, and preparing the way for a dramatically improved next generation of space stations. Everyone involved in these projects had much to learn. Engineers building Salyut and Skylab shared concerns about size, shape, weight, and strength. Protection from micrometeors, insulation from extreme temperatures, and how to attach many of the external components also consumed engineers from both nations.

Other teams focusing on the interior of space stations had concerns about how astronauts would live and perform their experiments. The interior space had to be designed to accommodate as many as three astronauts, along with supplies of food, water, and oxygen to last for many months. The interior space also had to accommodate occupants' need for sleep, exercise, hygiene, and personal privacy—this in addition to space for conducting several experiments involving a variety of scientific devices. Still others worked to solve problems involving the generation of electricity, keeping the interior spaces clean and safe, and taking into account unique problems that working in a weightless environment presented.

The Architecture of Space Stations

The architecture of both Salyut and Skylab were similar in that they both consisted of a single large cylinder within which crews lived and performed their experiments. Orbiting Earth at roughly 250 miles in space, these laboratories required relatively little structural reinforcement because their orbits placed them in a vacuum where pressure was negligible. Furthermore, although the craft would orbit Earth at about 17,000 miles per hour, there would be no friction of atmosphere to tear away external equipment. Secondary components necessary for sustaining the crew and supporting their scientific activities, therefore, were attached to the exterior surface of the spacecraft. As a result, both Skylab and Salyut would be anything but streamlined.

Other alterations to the basic cylindrical shape were also necessary. For example, a device known as a multiple docking adapter was needed that would firmly lock transport vehicles in place. To allow crew members to exit and reenter for space walks, technically called extravehicular activities, an airlock module, capable of accommodating one crew member at a time, was attached. Each space station would carry one or more externally mounted telescopes, to be used for photographing the Sun and other objects in the solar system. Externally mounted antennae would receive and send radio signals. And to provide the stations with electricity, multiple solar panels would reach outward. On Skylab, the four solar panels, each forty-eight feet long, would collectively look like the blades of an old-fashioned windmill.

Although both space stations were bulky and relatively heavy, one focus of their design was compactness. Both Salyut and Skylab were assembled on Earth, tightly packed into the cargo bay of a single rocket, and then blasted into orbit. To accomplish this feat, parts of the space stations were folded before insertion into the cargo bays. Then, when they were released in space, the folded parts were meant to unfold. Space aboard the launch vehicle was, therefore, at a premium. As a consequence neither the United States nor the Soviet Union sent crews on the rockets carrying the space stations. Instead, it was expected that crews would be sent once all components had safely deployed and the stations had stabilized in their respective orbits.

Optimizing Internal Space

Engineers realized that the space station envisioned was far too large to place in orbit with the boosters then available. Therefore, the interiors of Skylab as well as Salyut were designed to optimize precious usable space. The tons of scientific equipment, food, water, and personal effects of the crews meant that every cubic millimeter of space would be needed. Years prior to launch, when engineers were still designing the stations, they sketched to scale every item that would go aloft to ensure that it would actually fit. Since astronauts in a weightless environment occupy all three dimensions of a room as they float from place to place, artists familiar with human anatomy were employed to sketch in astronauts to ensure they would have adequate room to maneuver. Engineers also weighed every item to within the accuracy of one gram to verify that the fully loaded space station would not be too heavy for the rocket to lift it to the proper orbit.

Skylab, with an overall length of 118 feet and a diameter of 22 feet, had an interior volume of roughly 10,000 cubic feet, the equivalent of a small house. Divided into two stories, Skylab was designed for a work laboratory on the upper story, which occupied 38 percent of the interior, with the larger lower story for living quarters. This lower story was subdivided into a wardroom (used for dining and exercise), sleeping compartments, and an enclosure housing a shower/toilet.

Since interior space was at a premium, creative approaches to its use were encouraged. For example, aboard Skylab all tables and many other horizontal surfaces were designed to be folded up against the walls when not in use to make room for larger experimental apparatuses. Even the shower facility folded up. To conserve space, clothing for all astronauts was vacuum sealed, compressed, and stowed in special containers.

Engineering for Weightlesness

No anticipated condition occupied more time and required more engineering considerations than that of weightlessness. In designing interiors, engineers had to take into account that everything not secured in some way would be floating. In such an environment,

What Is Weightlessness?

Gravity is a force that governs motion throughout the universe. It holds all things to the ground, keeps the Moon in orbit around Earth, and Earth in orbit around the Sun. Many people mistakenly think that astronauts on the space stations are in a zero-gravity environment high above Earth and this is why they float throughout space stations along with their pens, paper, and anything else not tied down. Nothing could be further from the truth.

According to physicists who understand the mathematics of gravity, an apple falling from a tree on Earth would also fall to Earth if the tree were stationary 250 miles out in space, where most space stations orbit. The gravitational field is still quite strong there, roughly 95 percent of what it is on the surface. Why, then, do space stations not crash to Earth, and why do astronauts appear to be weightless in their cabins?

Weightlessness can be created two ways. One way is to travel millions of miles from any large object, where the gravitational pull diminishes to the point where it is very small. The second way, and a much more practical method, is to create a weightless environment through the act of free fall. Free fall occurs temporarily on Earth on roller coasters that crest and then suddenly drop down the rails or on airplanes that temporarily execute a steep dive. The same principle is used on space stations.

Physicists explain that just like the roller coaster and the diving plane, space stations are in a constant state of free fall. The key to maintaining them in a constant free-fall state is to keep the space stations traveling at just the right speed and right altitude—17,000 miles per hour and 250 miles up. Given these two conditions, space stations will perpetually fall around the revolving Earth; their orbits are curved and parallel to Earth's surface, not straight. As long as the speed and altitude remain constant, they will remain in a perpetual state of free fall.

On board the space station, the astronauts are also falling; it just does not look like they are. That is because they are falling along with the space station. Since they are falling at the same rate as the space station, which is constant free fall, they appear to float in the state that physicists call weightlessness.

astronauts could not work effectively if they had to contend with floating objects. To control the interior, designers fabricated special tie-downs for commonly used objects, some wall space was lined with Velcro to which tools and other objects could be quickly attached, and all containers had to have covers to prevent objects from floating away.

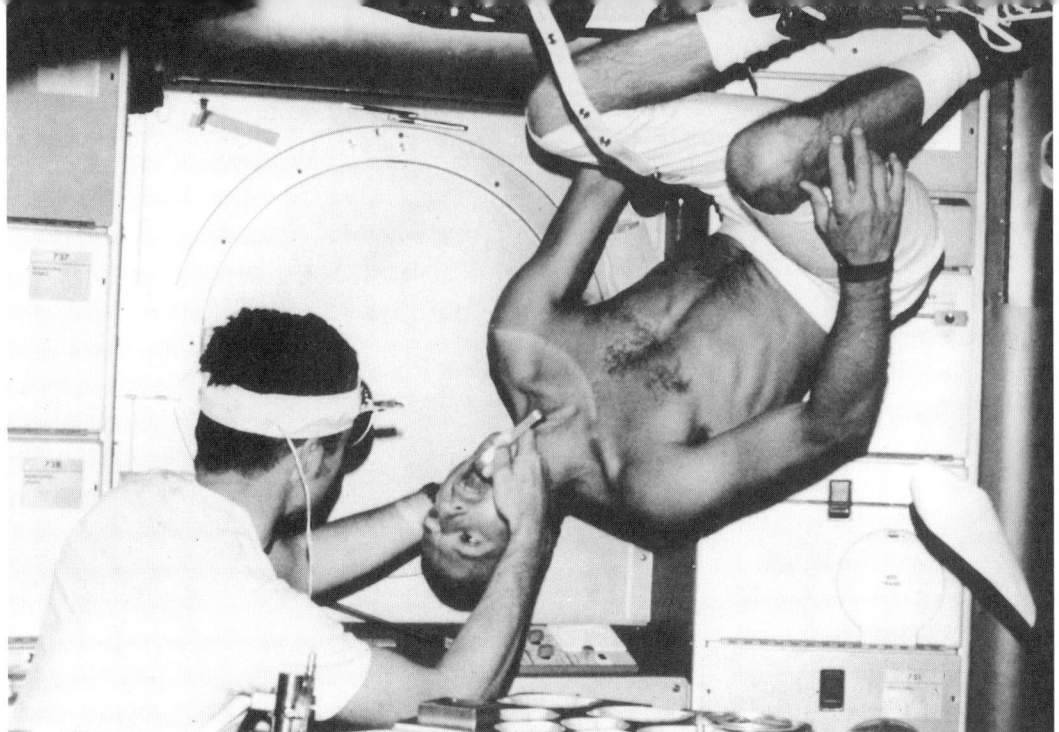

Even routine activities like a medical exam are a challenge in a weightless environment like that found on the space station Skylab.

Designers had to anticipate that the contents of containers would behave differently in orbit than on Earth. Carbonated soft drinks, for example, are particularly problematic because in a weightless environment bubbles of carbon dioxide remain randomly distributed in the fluid rather the rising to the top. The result can be a foamy mess when an ordinary can is opened. By dispensing carbonated soft drinks in a collapsible bag instead of a bottle, however, the pressure around the fluid can be constantly controlled, preventing a messy explosion.

The engineers also had to design interiors with weightlessness in mind. Moving from compartment to compartment through tight hatches or down narrow corridors while weightless meant that a strong push from a leg or arm might send an astronaut colliding with walls. To lessen the dangers of bumps, bruises, or worse, hatches were padded, as were the walls of corridors. To assist crew members moving about and to help them stay put when they wished to do so, some walls and ceilings were fitted with a metal triangular grid work. Shoes designed for the astronauts had triangular plates fastened to the soles that would fit through the triangular opening in the grid. By turning one foot slightly, an astronaut would be able to hold steady while working.

There were also other means of restraint. One was a set of straps similar to those on beach sandals. Three pairs of these were placed on the floor of the wardroom at the base of the food table; another pair was located in the waste management compartment, where garbage was collected, sorted, and sealed; and yet another pair was installed in front of the toilet. Still another type of foot restraint, for use with the heavy boots of space suits, consisted of toe bars and heel fittings which would be fastened to floors and walls of the airlock that astronauts used before exiting on space walks.

Performing the many scientific experiments while weightless also presented unique problems. Some experiments required the assembly of an apparatus involving hundreds of small parts such as tiny screws, nuts, and bolts. If a container of small parts was accidentally shaken or bumped, setting in motion dozens of small parts, progress came to a halt until all the parts could be rounded up and counted. In time, astronauts on both space stations learned that air currents, created by ventilation fans within the space stations would carry most lost objects to a filter, where they could easily be retrieved.

A greater hazard to the safety of crews were liquids lost and floating throughout the space stations. When liquids got loose from their containers, they often broke into thousands of tiny droplets, which then dispersed throughout the interior. Although each space station had small devices for vacuuming up liquid spills, there was always the danger that a liquid might slip into electrical wiring and short out critical components.

Powering the Space Stations

In an outpost orbiting more than two hundred miles above Earth, the one critical commodity needed in large supply that could not be packaged and flown into orbit was electricity. Although large batteries were once considered, they were quickly rejected for long-term flights because their excessive size and weight made it impractical to send them into orbit. Instead, engineers worked to utilize and improve upon the existing technology for converting solar energy into electricity, a process known as photovoltaic generation.

The principles of converting light to electricity had long been known, and the technology for manufacturing photovoltaic cells had been developed by the 1950s. Although they were relatively inefficient at that time, by the 1970s photovoltaic cells themselves had improved significantly, as had wiring, which significantly reduced loss of electrical current. Because of these improvements, the problem of supplying electricity to space stations was solved.

In fact, the most distinctive external feature of both Salyut and Skylab was their array of solar pan-

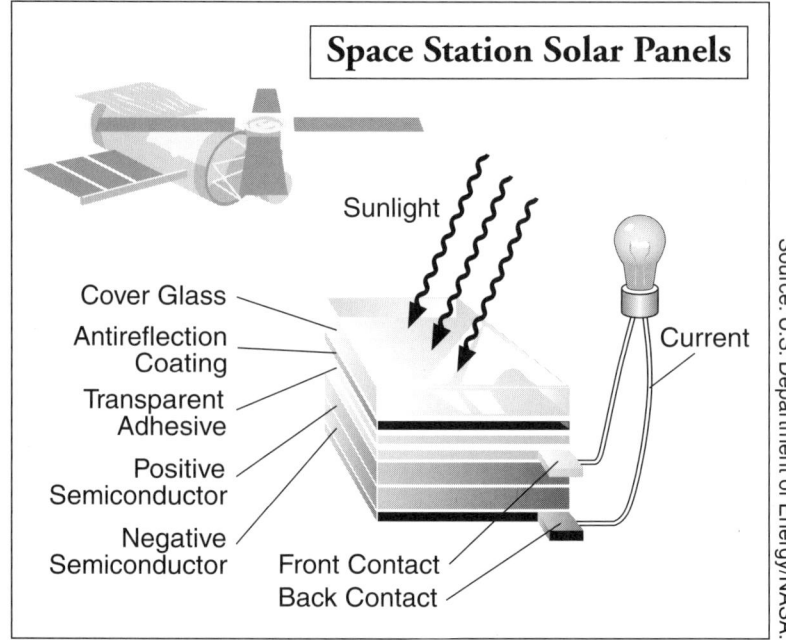

els that stretched outward to capture the Sun's rays. These panels consisted of thousands of thumb-sized silicone chips that directly converted sunlight into electricity. Both Skylab and Salyut deployed about one thousand square feet of panels to produce the 120 volts of direct current needed to run the hundreds of small motors, compressors, lights, computers, refrigerators, and navigation and communications systems.

Since sunlight was the necessary ingredient, electrical generation could only occur when the photovoltaic panels were bathed in sunlight. Onboard computers controlled the panels' orientation by firing small rockets to slightly change the orientation of the spacecraft to maximize exposure to sunlight yet avoid overheating the interior of the spacecraft. Since all space stations circle the globe sixteen times a day, there were sixteen times daily when Earth obstructs sunlight. During those dark moments, space stations would have to rely on batteries to supply power.

Learning from Mistakes

In spite of unbridled enthusiasm shared by the United States and the Soviet Union for their first attempts to establish space stations, initial attempts were fraught with problems. No one involved was especially surprised because everyone understood that the technical leaps being made were large and that many of them were untested.

The first test of space station technology was Salyut 1. Two days after Salyut 1 was sent into orbit, on April 23, 1971, the transport vehicle called Soyuz 10 docked to deliver the first three Soviet cosmonauts. As the three saw their space station coming into view, they maneuvered the nose of their vehicle into a five-foot-diameter docking port to establish the initial "soft" docking. The two craft coupled flawlessly, but when the cosmonauts then attempted the "hard" docking to secure and hermetically seal the two vehicles together with interlocking steel collars, the locking mechanisms failed to properly mate. Following six hours of frustration high above the earth, the cosmonauts withdrew from Salyut 1 and returned home later that same day.

Undaunted by what the Soviet space agency described as a partial success, engineers made mechanical adjustments to the steel collar and a second attempt to dock with Salyut 1 took place on June 6 of the same year. This time the docking was successful, and television cameras transmitted pictures of the first men to inhabit a space station turning somersaults as they entered their new home. The schedule called for the men to inhabit the station for thirty days while performing a variety of experiments, undock on July 6, and return home.

Over the next eleven days, life on board Salyut had become a predictable routine, and most observers believed that the mission would flawlessly continue. Suddenly, however, on June 18 a small electrical fire in one of the thousands of electrical cables broke out. The three men retreated through the docking port to their return vehicle and requested permission to terminate their mission and return home. Ground command, however, provided directions for extinguishing the fire, and the men returned to Salyut and successfully performed the task. Unfortunately, this unforeseen event shook the men's confidence, impairing their ability to complete the thirty-day mission. On June 29 ground control made the decision to bring the three men home ahead of schedule. The television cameras recording their departure captured their relief to be returning home as they laughed and joked about the electrical fire that had at one time placed their lives in peril.

At this point the cosmonauts' luck took a fatal turn. The computer systems on board Soyuz 10 performed flawlessly. Following reentry into the atmos-

Launched April 23, 1971, the Soviet Union's Salyut I was the first space station to orbit Earth. Soviet cosmonauts docked with the station two days later.

phere, the speed of Soyuz 10 dropped from seventeen thousand miles per hour to three hundred miles per hour by the time the vehicle was at an altitude of ten thousand feet. At that point the three parachutes deployed and softly deposited the capsule back on Earth at the precise predetermined spot. The only hint that something might be wrong was an unexpected loss of communications. The recovery team gathered around the capsule and unbolted the hatch to give the three cosmonauts a heroes' welcome, but it was shocked to discover all three silently sitting motionless upright in their seats. All three were dead.

An inquiry into the tragedy revealed that a hatch valve was either opened or jolted open when the descent vehicle separated from Salyut. The open valve allowed the precious supply of oxygen to escape into space. The valve would normally have opened as the spacecraft descended through the lower levels of the atmosphere, equalizing the pressure inside and outside of the spacecraft. Opening prematurely in space, however, proved fatal.

Improvisation on Skylab

American engineers, as they completed preparations for the launch of Skylab, studied the fatal failure of Salyut and paid careful attention to its many problems. Nonetheless, Skylab's early history was a rocky one, although no astronauts lost their lives. On May 14, 1973, the unmanned Skylab station was catapulted into space on its maiden voyage by a Saturn 5 rocket. All systems performed flawlessly until sixty-three seconds into the launch, when controllers received an unexpected signal from Skylab indicating that the station's critical micrometeoroid shield had deployed—while still stored inside the rocket ship. Apparently vibrations during launch

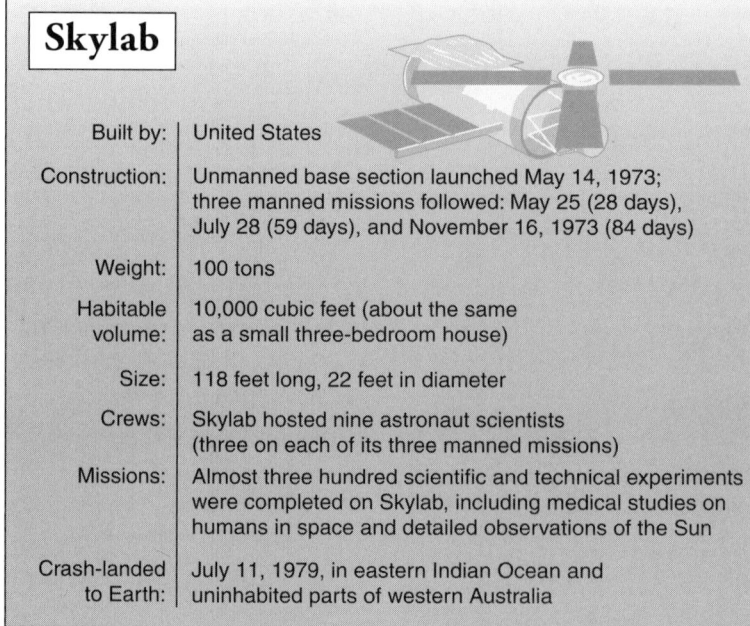

Skylab

Built by:	United States
Construction:	Unmanned base section launched May 14, 1973; three manned missions followed: May 25 (28 days), July 28 (59 days), and November 16, 1973 (84 days)
Weight:	100 tons
Habitable volume:	10,000 cubic feet (about the same as a small three-bedroom house)
Size:	118 feet long, 22 feet in diameter
Crews:	Skylab hosted nine astronaut scientists (three on each of its three manned missions)
Missions:	Almost three hundred scientific and technical experiments were completed on Skylab, including medical studies on humans in space and detailed observations of the Sun
Crash-landed to Earth:	July 11, 1979, in eastern Indian Ocean and uninhabited parts of western Australia

Source: NASA (http://science.nasa.gov).

had caused the delicate shield to tear and prematurely deploy.

The micrometeoroid shield was a critical component for two reasons. First, it was intended to protect Skylab's workshop wall against penetration by micrometeoroids speeding through space. Such tiny objects were common, so the danger they presented was real. Secondarily, the shield was designed to provide shade while Skylab was exposed to the direct rays of the Sun. Without the shield, the space station would rapidly overheat.

Of equal concern was the damage the shield did to other components when it deployed. When Skylab was released from the rocket bay and allowed to drift free, engineers on the ground discovered the true extent of the problem. When the shield opened prematurely, it not only tore, but it also ripped away one of the craft's solar panels. With only three solar panels left, ground engineers needed to maneuver Skylab to face the Sun so as to generate as much electricity as possible. Because of the loss of the micrometeoroid shield, however, this repositioning caused internal temperatures to rise to 160 degrees Fahrenheit—far too hot for humans to endure.

The crew remained on the ground; meanwhile, Skylab was in a crisis. On the one hand, excessive heating could spoil food and film and could cause insulation to give off poisonous fumes. Lack of electricity, on the other hand, would cripple the workshop and other critical electronic equipment such as computers and navigation systems. Without a resolution, ground crews would be forced to scuttle the entire project.

James Kingsbury, deputy director of the Astronautics Lab, told the team to troubleshoot the problem and to "keep the vehicle in a mode where we can inhabit it and find out a way to fix it. Whatever you need at the center is yours. This is the one thing we are going to do at the moment. We will turn on

Skylab orbits Earth with the metallic parasol (bottom) that replaced the damaged micrometeoroid shield. The space station fell out of orbit in 1979 after three missions.

everything and everybody we have who can do anything."⁶ The agreed-upon solution was to maneuver Skylab by remote so that its working solar panels were at a 45-degree angle to the Sun. This compromise reduced interior temperatures to a safer 122 degrees Fahrenheit but still generated some electrical current.

On May 25 the three-man crew went up to begin repairs. One task involved deploying a makeshift metallic parasol, after which the temperature dropped dramatically. Now Skylab could be turned to expose the solar panels to more of the Sun's rays to increase electrical generation. These successful repairs allowed the crew to spend twenty-eight days at the station followed by two subsequent crews for an additional fifty-nine and eighty-four days, respectively by the end of 1973.

The results of the troubleshooting by engineers on the ground and improvisations by astronauts on Skylab produced 171 days of operation that satisfied the American engineers who designed it and the scientists who had an opportunity to conduct experiments. A total of 117,000 photographs were taken of the Sun, showing extraordinary details of the solar surface, close-ups of sunspots, and the violent solar winds that can sweep across the massive hot solar surface. The store of information captured in space exceeded what had been amassed in the previous hundred years of study from the earth's surface.

Earth had also been the subject of constant observation, resulting in forty-six thousand photographs. Meteorologists, oceanographers, and biologists scrutinized every one for any detail that might add new information about the general health of the planet and any signs of environmental stress. Added to these photographs were additional databases of information, the results of hundreds of hours of experimentation in material sciences, biology, and space medicine. Above all, the study of the response of astronauts' bodies, to prolonged periods in a weightless environment met one of the most important objectives of the Skylab program, simply to confirm that humans could survive in space for prolonged periods.

Skylab fell from orbit in 1979 following just three missions, totaling 171 manned days. Salyut 1, meanwhile, was followed by a succession of six space station launches, each bearing the Salyut name. The combined lessons learned were of inestimable value. Most of all, confidence in the combined problem-solving capability of ground crews and astronauts was sufficient to move both the United States and the Soviet Union forward toward more ambitious missions requiring more sophisticated space stations.

Chapter 2

A Quantum Leap in Technology

The Skylab and Salyut missions firmly established the feasibility of extended stays in orbit and of performing fundamental—if rudimentary—research. In order to move beyond their limitations, a quantum leap forward in technology, something truly revolutionary, would be required. As former Houston Skylab program manager Robert F. Thompson observed, Skylab was a

> beautiful tactical program that had numerous shortcomings as a strategic program. Skylab was not designed for in-flight repair, re-supply with air and water, refurbishment with improved technology, re-visitation for re-boost to a higher orbit, or restructuring as part of a larger station. Consequently it could not, and did not, lead to a strategic, sustained human presence in space.[7]

Although America's space agency, the National Aeronautics and Space Administration (NASA), chose not to follow up on its limited space station success, opting instead to develop a reusable space vehicle, the shuttle, the Soviets opted for a different course. Despite a horrific beginning for their Salyut program back in 1971, many successes followed. By the time the Salyut 6 and Salyut 7 projects had ended in the late 1980s, the Soviets' many discoveries decisively

A Quantum Leap in Technology

eclipsed their failures while at the same time eclipsing the more modest successes of Skylab.

The Salyut program's many accomplishments opened the way to the creation of the Soviets' Mir space station; this constituted the quantum leap aerospace engineers such as Thompson had asserted would be necessary if a sustained human presence in space was to be possible.

Mir—the Beneficiary of Lessons Learned

During the late 1970s, the Soviet Union committed itself full-bore to placing a space station in orbit that would vastly surpass the Salyut and Skylab

The space shuttle Atlantis *docks at Mir space station in 1993. Soviet engineers designed Mir to be far superior to the earlier Salyut and Skylab stations.*

space stations. *Mir,* a Russian word meaning "peace," drawing on lessons from both the successes and failures of Salyut, was primarily intended to prove that humans could survive in a weightless environment for years, not just months. Secondarily, Mir would also be called upon to serve as a laboratory where more complex experiments than those conducted on Skylab and Salyut could be done.

Many improvements would be needed to meet Mir's ambitious objectives. Highest on the list was a significantly larger interior space consisting of multiple compartments in which many different types of experiments could be conducted simultaneously. Also high on the list were improvements to the docking mechanisms so transport vehicles could more readily resupply the crews, improvements to the electrical systems, more sophisticated equipment that allowed for lengthier space walks, and more sophisticated optical equipment for viewing Earth and galaxies beyond the Milky Way.

Modular Orbital Assembly

Soviet scientists knew that a larger space station would make extended stays in orbit practical. The problem, though, was that a space station of the size they envisioned exceeded the weight and volume limits imposed by the Soviets' Proton rocket. In response, engineers designed Mir in six separate modules, each of which would be flown to orbit separately and then joined 250 miles above Earth. Some engineers describe this sort of module construction as "tinker toy" construction, but others liken it to snapping together Lego blocks.

Taken together, Mir's six modules would be roughly twice the size of Skylab and would collectively function as crew quarters, a computerized brain and communications center, and research facilities. The first of the six modules sent up, the Mir Core, was launched on February 20, 1986. As its name implied, the Core was engineered with six docking ports to which the other modules later would be attached. The Core was also aptly named in that it contained both the basic living quarters and research laboratories. The basic living quarters included individual crew quarters, a bathroom, shower facilities, and a small galley with cooking facilities and a table. The Core, therefore, was capable of functioning as a stand-alone space station until the additional modules could be sent up.

Following the successful deployment of the Mir Core, the remaining five modules were sent up and locked into place over a four-year period. In all cases, each module was loaded into the transfer vehicle, Soyuz, which was in turn loaded onto the Proton rocket and was launched into space. Once in space, the Proton rocket separated from Soyuz, which was then able to rendezvous with Mir by firing its thrust rockets. Once docked with the Core, cosmonauts were then able to remove each module and send Soyuz back to Earth for future reuse.

Coupling each module to the Core's docking ports required space walks on the part of the cosmonauts to secure the locking mechanisms, pressure seals, and electrical wiring. Despite many mechanical improvements, coupling the modules proved more complicated than anticipated. For example, when the Kvant module, which provided data and observations of active galaxies, quasars, and neutron stars as well as some biotechnology experiments, approached the Mir Core in 1987, something very wrong occurred. As it set up to rendezvous with its dock, the horrified crew on the Core watched in disbelief as the twenty-ton module floated past too high to lock. Four days later, a second approach went according to plan, but the final locking mechanism jammed. Forced to take another space walk to investigate the problem, the cosmonauts discovered a plastic trash bag lodged in the locking collar. Once removed, the Kvant module successfully locked in place.

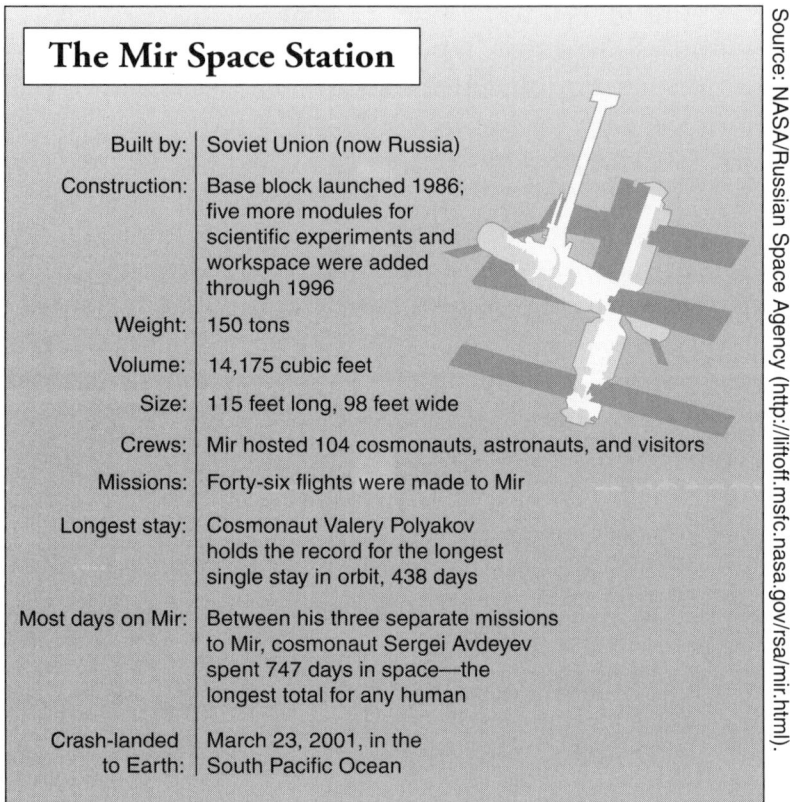

The Mir Space Station

Built by:	Soviet Union (now Russia)
Construction:	Base block launched 1986; five more modules for scientific experiments and workspace were added through 1996
Weight:	150 tons
Volume:	14,175 cubic feet
Size:	115 feet long, 98 feet wide
Crews:	Mir hosted 104 cosmonauts, astronauts, and visitors
Missions:	Forty-six flights were made to Mir
Longest stay:	Cosmonaut Valery Polyakov holds the record for the longest single stay in orbit, 438 days
Most days on Mir:	Between his three separate missions to Mir, cosmonaut Sergei Avdeyev spent 747 days in space—the longest total for any human
Crash-landed to Earth:	March 23, 2001, in the South Pacific Ocean

Source: NASA/Russian Space Agency (http://liftoff.msfc.nasa.gov/rsa/mir.html).

Once all modules were properly locked onto the Core, Mir weighed about 150 tons, with dimensions of roughly 115 feet by 98 feet and an inner volume of about 14,100 cubic feet, significantly larger than Skylab had been. Orbiting high above Earth, the fully assembled Mir was the biggest man-made object in space.

Resupply

Another of Mir's revolutionary improvements over previous space stations was its ability to support crews for long periods by allowing for delivery of fresh supplies. When the Soviets developed Mir, they also developed for it an unmanned supply vehicle named "Progress." To accommodate Progress for quick and reliable docking, Mir's docking port was redesigned with an improved locking assembly and with a computer program to automate docking.

The new locking assembly was a set of four hydraulic grappling hooks capable of extending ten inches to capture Progress's port and then retracting to pull Progress directly into Mir's docking port. Assisting in this mechanical ballet was a computer program on board Mir capable of sensing the position of Progress and Mir and then adjusting the orientation of each for flawless dockings.

Progress was basically a cylinder with an interior storage volume of twenty-five hundred cubic feet, enough to carry sixteen thousand pounds of supplies. Since Progress was relatively small, one small rocket could propel one supply vehicle to Mir; alternatively, a larger Soyuz SL-4 rocket could propel several supply vehicles. Because of this ability to resupply Mir, the Soviets would later set the record for the longest stay by a human in space, 438 days.

Progress delivered everything needed aboard Mir for experiments and for the crew's use. Most of Progress's cargo consisted of propellants and gases such as hydrogen, helium, argon, and oxygen, needed for experiments. In addition to propellants and gases, Progress ferried numerous housekeeping items to the station. The most vital part of this package for the crew was life-support hardware and supplies, such as chemicals that release oxygen and others that remove carbon dioxide from Mir's atmosphere. Other items included such necessities as computers, communication equipment, and expendable hardware such as bolts and electrical wire. Of even greater interest to the crew were personal items such as toothbrushes, toothpaste, combs, brushes, medical kits, laptop computers, and pens and pencils. Food and water were, of course, necessities, but unlike in the case of Salyut, Russian officials made certain that the food was not just nutritious but also palatable. According to David M. Harland, author of *The Mir Space Station: A Precursor to Space Colonization*,

"Much of the food was fresh, and specialties such as apples, onions, garlic, and caviar were very much appreciated."[8]

Mir fulfilled all of its expectations. It would remain in orbit for fifteen years, during which time it was never unoccupied. Four different individuals would each spend more than twelve months on Mir. Harland notes, "The Mir complex is a tentative first step towards an orbital habitat. This, along with the evaluation of the human organism in a weightless state, is Mir's *raison d'être* [reason for being]. Mir has succeeded in its mission."[9]

As Mir began to age, a debate arose among Soviet scientists over refurbishing it or scrapping it in favor of a new-generation space station. Finding the money for such a project was a problem, however, so the Soviet Union explored enlisting the financial and engineering assistance of other nations. During the late 1980s and early 1990s, an easing of geopolitical tensions between the Soviet Union and the United States and its allies made this idea practical. Furthermore, unlike the 1950s, 1960s, and 1970s, when space was exclusively the competitive arena of the United States and the Soviet Union, now other technologically advanced nations were willing to participate in and benefit from an orbiting outpost.

The International Space Station

American space scientists had opted to develop the space shuttle rather than compete with Mir by developing another space station. Still, in 1993 President Bill Clinton proposed to Russia and several other nations a cooperative effort to place the next generation of space stations in orbit. This novel and timely idea of sharing the costs, the risks, the technology, and the results of research spurred ambitious designs for the International Space Station (ISS). If all went well, the first of more than one hundred components of the ISS would be in orbit in 1998. Plans

The Death of a Space Station

Relegating space stations to the junk heap is both a tricky and complicated proposition. All space stations will eventually fall out of orbit and plummet back to Earth—unless regularly propelled back into orbit—as their speed and altitude drop due to Earth's gravitational pull. In 1991 Russian aerospace engineers finally allowed Salyut 7 to fall out of orbit and plummet to Earth. Unfortunately, Russian space engineers had no idea where it would hit the earth, and even after disintegrating as it plunged through the atmosphere, several tons of debris eventually scattered across the Andes mountains, much to the anger of many people.

Hoping to avoid another international incident, the Russians planned better when the time came to bring Mir out of orbit. In March 2001 the world was notified that the 130-ton Mir would be brought down somewhere over the Pacific Ocean, the largest unpopulated region on Earth. As the space station gradually lost altitude, the date of splashdown in the Pacific was announced as March 21. Soviet engineers announced that as it passed through the atmosphere at initial speeds of seventeen thousand miles per hour, 110 tons of the 130-ton craft would burn up from friction; the surviving 20 tons would scatter into thousands of small pieces that would splash into the ocean.

Hitting the Pacific on a particular day would require control. Soviet engineers conceived the idea of actually accelerating Mir's descent through the atmosphere to control the time and place of disintegration. To accomplish this objective, a Progress cargo ship was attached to the station, which had a rocket propulsion system. When Mir fell to about 136 miles, the rocket engines on Progress would be fired to accelerate Mir downward into the thicker layers of the atmosphere, where it would quickly break apart and burn.

The controlled reentry was projected to bring the craft down in a 380,000-square-mile swath of the Pacific between New Zealand and Chile, away from major air and sea routes. Tiny nations throughout the South Pacific were alerted to watch for the chunks of Mir, and dozens of island authorities warned their people not to go out March 21 and to stay off boats to avoid being hit by any parts.

In a trip condemned as suicidal by Russia's space agency, a California-based public relations firm chartered an airplane for a group of space enthusiasts and television crews to fly to the site. They hoped to photograph the blazing reentry, but NASA authorities estimated their chances of being hit as about 1 in 2 billion. Fortunately for everyone, the plan worked and the blazing scraps of Mir plunged harmlessly into the Pacific.

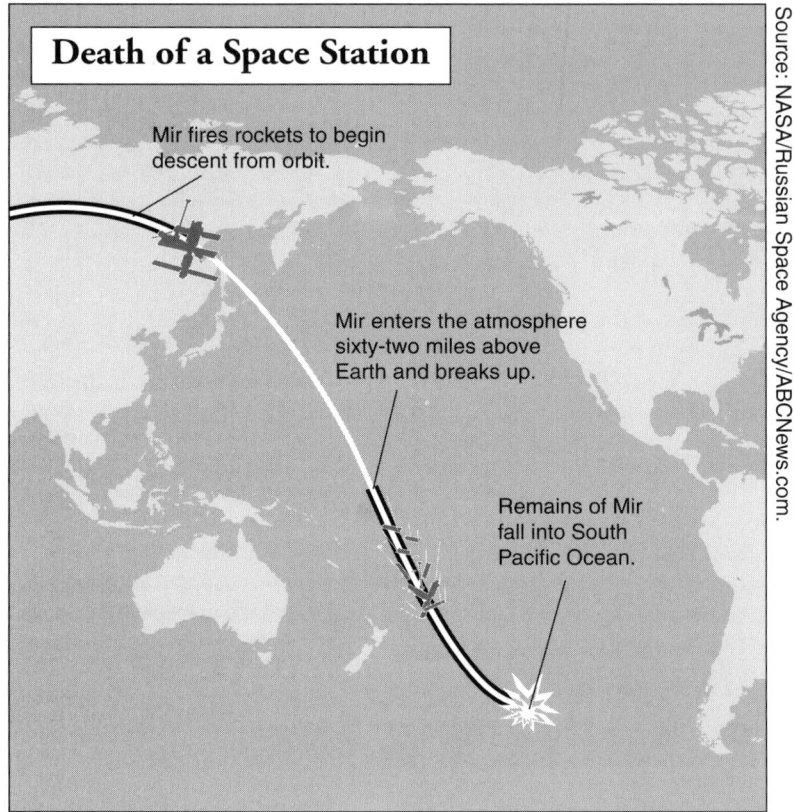

called for the locking of the remaining modules by the end of 2003, with experiments beginning immediately and running continuously until at least 2013.

Everyone involved in the project understood that the ISS would be the culmination of thirty years of research and experimentation on Salyut, Skylab, and Mir. Far from being just the next iteration of space stations, aeronautical engineers worked to make the ISS a second quantum leap in the wake of its predecessor, Mir. According to Daniel Goldin, the head of NASA, the ISS would push space station technology far beyond Mir:

> The International Space Station (ISS) will change the course of human history. The ISS is certainly an ambitious idea. It is probably the largest international scientific and technological

A Quantum Leap in Technology

project ever undertaken. The goal is to establish and maintain a permanent presence in space and to provide a testbed for new technologies, medical research, and the development of advanced industrial materials.[10]

The design of the ISS would be on a scale that would dwarf Mir. Rather than the total of five modules that Mir had, designers planned for six primary modules used exclusively as laboratories in addition to additional modules for crew quarters, storage facilities, docks for transport vehicles, and airlocks needed for space walks. All of these modules, when coupled with six pairs of solar panels and external apparatuses such as telescopes and communications antennae would give the ISS the look of an oversized spider moving through space.

An artist's conception of the completed International Space Station shows its huge, spidery form. Five international space agencies are collaborating on the project.

Significantly larger than Mir, the ISS structure would have an overall length of 262 feet, a width of 365 feet, and a total weight of a massive 500 tons, nearly four times the weight of Mir. NASA engineers liken the total exterior space to the dimensions of two football fields and the interior space to that of the passenger cabin of a Boeing 747. Equally remarkable will be the ability of the ISS to accommodate a crew of seven—more than twice the capacity of Mir.

International Cooperation

The first step toward building the ISS was assigning responsibilities and costs to each of the charter nations. In 1993 most of the details of the joint space contract were ironed out and agreed upon. In his book *Living in Space: From Science Fiction to the International Space Station*, space writer Giovanni Caprara comments on this agreement:

> The signing of the joint space agreement marked the end of an era of antagonism [between Americans and Russians] and the beginning of a new phase of cooperation. . . . Until now, space programs had been viewed as an ideal means of demonstrating the superiority of a political system. Now they became a proving ground for experiments in cooperative agreements that could be usefully applied to other fields.[11]

The United States and Russia were joined in the construction effort by a consortium of ten European nations—called the European Space Agency (ESA)—as well as by Canada and Japan. Engineers realized that overcoming the forty years of competition between the United States and Russia would be an asset to the new space station. Canada declared its intention to construct the robotic arm that would be used for assembly of the ISS modules and for placement and retrieval of a variety of equipment for experiments. Fifty-seven feet long, 15 inches in diameter, and weighing 911 pounds, the robotic arm derives its flexibility from six revolving joints and its grasping ability from pincers designed to maneuver a 266-ton object while in orbit. Since the robotic arm would be an essential component for assembly of the ISS, it would be one of the first items flown to the station.

Japan announced its interest in building a module called the Japanese Experiment Module (JEM), intended to be a multipurpose facility for a variety of space science and technology studies. Nicknamed "Kibo," the JEM is a cylinder thirty-seven feet long and ten feet in diameter. Attached to Kibo is an external platform similar to a back deck, called the Exposed Facility, available as a storage unit and laboratory for conducting experiments intended to be performed in a vacuum.

The ESA, which consisted of three major contributors—France, Germany, and Italy—and seven minor ones, agreed to contribute a research module called the Columbus Orbital Facility and a transfer vehicle that would be used to transport supplies to the ISS and to boost the orbit of the station to a higher altitude if needed.

Russia, the country with the most experience in long-term missions on space stations, was called upon to make a considerable contribution to the ISS. Russia agreed to build the first module that would go into orbit, the FGB, which is the Russian acronym for Functional Cargo Block. It would function as the control center for the ISS, providing docking ports, fuel tanks, and solar panels. Weighing nineteen tons, this forty-foot-by-twelve-foot cylinder would be the largest of the ISS modules. The Russians also agreed to supply at least two science modules, additional solar panels, resupply vehicles, and, of critical importance, an escape vehicle to be used in the event of some catastrophe on the ISS that would necessitate the crew's evacuation. Russia's one last major contribution, the Proton rocket, would muscle the main pieces into space, requiring an estimated ninety launches.

America's role began with building Node 1, named "Unity," which would function as conduit for power, liquids, gases, and communications needed by all the other modules. Next, the United States built the American Laboratory, which was intended to be used for a variety of experiments; the United States would also supply eight solar panels. America agreed to provide the space shuttle, which, along with the Proton rockets, would fly the ISS modules into orbit and provision them over the life of the space station.

Leading-Edge Technologies

ISS engineers applied the latest leading-edge technologies to create a safe interior work space for the crew. Each module has an outer shell of lightweight aluminum. This shell has an additional protective layer of four-inch-thick impact-resistant Kevlar and ceramic material. This layer functions as a bulletproof vest to provide extra protection from impacts by micrometeoroids and tiny grains of grit that punctured previous space stations, causing air leaks.

Another significant departure from Mir's design is the high degree of specialization of each ISS module. This specialization of function necessitates a great deal of interdependence with other modules. Unlike Mir, on which each module could function independently of the others, none of the ISS modules can survive in space without the assistance of the others. In this regard, aerospace engineers liken the ISS design to the human body, in which each of the body's systems has a distinct and highly specialized function yet each is dependent upon the proper functioning of all others.

Such a design will make possible more complex experiments that will answer more questions about

A Quantum Leap in Technology

International Space Station Modules

Russian Sections
- **R1** Universal Docking Module
- **R2** Docking and Stowage Module
- **R3** Research Module (2)
- **R4** Docking Compartment*
- **R5** Zvezda (Star) Service Module*
- **R6** Soyuz*
- **R7** Science Power Platform

Japanese Sections
- **J1** Kibo (Hope) Japanese Experiment Module (JEM)
- **J2** JEM Pressurized Section
- **J3** JEM Pressurized Module
- **J4** JEM Remote Manipulator System
- **J5** JEM Exposed Facility
- **J6** JEM Exposed Section

European Sections
- **E1** European Lab Columbus Orbital Facility

Brazilian Sections
- **B1** Express Pallet

Canadian Sections
- **C1** Mobile Remote Servicer Base System*
- **C2** Canadarm 2*
- **C3** Special Purpose Dexterous Manipulator

United States Sections
- **U1** S0 Truss Segment*
- **U2** S1 Truss Segment*
- **U3** Solar Alpha Rotary Joint
- **U4** S3 Truss Segment
- **U5** S4 Truss Segment
- **U6** S5 Truss Segment
- **U7** S6 Truss Segment
- **U8** Z1 Truss Segment*
- **U9** P1 Truss Segment*
- **U10** Solar Alpha Rotary Joint
- **U11** P3 Truss Segment
- **U12** P4 Truss Segment
- **U13** P5 Truss Segment
- **U14** P6 Truss Segment*
- **U15** Unity (Node 1)*
- **U16** Node 2
- **U17** Pressurized Mating Adapter 1*
- **U18** Pressurized Mating Adapter 2*
- **U19** Pressurized Mating Adapter 3*
- **U20** Quest Airlock*
- **U21** Cupola
- **U22** Zarya (Dawn) Control Module (Russian built / U.S. owned)*
- **U23** U.S Lab Destiny*
- **U24** Centrifuge Accommodation Module

* = Elements in orbit as of spring 2004

Source: NASA.

An Example of Complexity on the ISS

The engineering of each of the ISS modules is remarkably complex. An example is the eighteen-by-fifteen-foot cylinder built by the United States called Node 1, or the Unity Node, which functions as a conduit for all the essential elements of the ISS.

The aluminum walls of Unity are made of high-strength aluminum to withstand the impacts of delivery vehicles docking with it and the stress and torque of six other modules attaching to its six ports. To accomplish all of the requirements imposed on Unity by aerospace engineers, the American-made module contains more than fifty thousand mechanical items mounted in a maze of racks and tresses. This dazzling array of parts provides essential space station resources such as fluids, environmental control and life support systems, and electrical and data systems to the work and living modules.

To provide essential life support resources, Unity is packed with 216 pressurized conduits that carry fluids and gases. Since the weightless environment does not have natural air currents or liquid flow, gases, including oxygen, along with many liquids must be pumped throughout all human-occupied areas. And even more remarkable, this eighteen-foot-long module threads 121 separate internal and external electrical cables, which collectively consist of six miles of electrical and data wires.

American engineers designed the complex Unity Node, the conduit for all essential elements of the ISS.

the nature of physics and living in a weightless environment. According to Mary F. Musgrave, the associate dean of the College of Natural Sciences and Mathematics and a professor of biology at the University of Massachusetts, "The Mir and Skylab programs provided only a glimpse [of a space station's potential]. The International Space Station offers the opportunity to conduct research 24 hours a day, 365 days a year."[12] This view is echoed by many, including Al Feinberg, NASA public affairs officer at the Office of Space Flight, who added, "ISS is a radical departure from Mir."[13]

The Robotic Arm

The orbital assembly of the ISS began a new era of hands-on work in space. Although Mir was also a modular space station requiring orbital assembly, its design made the task quick and simple, involving not much more than clamps and cable hookups. The ISS would not be so simple. Because of the number of modules, their different configurations, and the unique requirements for scientific research, securing each one was a concern to designers from the start.

ISS engineers rejected the idea of clamps and cables in favor of mounting a large robotic arm, remotely controlled by crews inside the ISS, capable of seizing modules and coupling them with surgical precision. Built by the Canadians, the fifty-seven-foot-long robotic arm was one of the first components flown to and placed on the ISS. This arm, controlled by astronauts, will eventually assemble the full complement of the ISS modules. Once all modules are in place, the arm will also play a major role in supplying the space station, effecting repairs, and assisting in many experiments conducted outside of the pressurized environment of the station's modules.

What makes the arm work so effectively are seven motorized joints. These joints function much like wrists and elbows, only better. They are capable of revolving 360 degrees while arm segments extend in multiple directions. This arm is capable of handling large payloads and assisting with docking the space shuttle. To access all parts of the space station, the arm moves along rails running the length of the station and bolted to the truss framework.

Grasping is performed by an ingenious robotic "hand" called the Special Purpose Dexterous Manipulator, or Dextre for short. This device is capable of handling the delicate assembly tasks previously performed by astronauts during space walks. Equipped with lights, cameras, and tool packs, Dextre is capable of installing and removing small external payloads such as batteries, power supplies, and computers as well as manipulating, installing, removing, and inspecting scientific payloads. A typical task for Dextre is to replace a depleted 220 pound battery that involves bolting and unbolting operations as well as precision positioning to properly align and insert the spare battery within its work space and properly reattach all connectors.

The Escape Vehicle

Frightening moments on Salyut and Mir created life-threatening situations for cosmonauts on more than one occasion. Delicate mixtures of gases and chemicals needed for life support and scientific experiments created explosive situations from time to time. Added to this were fears of fire and catastrophic loss of atmospheric pressure should the skin of a module be punctured. To lessen the chances of a disaster on the ISS killing all on board, planners and engineers included in the design a permanent module that would function as an escape vehicle if needed.

Engineers designed the ISS to accommodate an escape vehicle permanently docked while astronauts worked in the station. The only available vehicle able to fulfill the requirements was the Russians' Soyuz transport vehicle that was designed to deliver cosmonauts to Mir and return them home. Unlike all other modules, Soyuz could be boarded and quickly released from the ISS in less than ten minutes.

Astronaut Buzz Aldrin, however, in an interview in *Popular Mechanics* magazine in 2003, pointed out one shortcoming of Soyuz as an escape vehicle, noting, "Currently the station [ISS] is limited to three, the number of people that can escape in the Soyuz capsule."[14] If the ISS is to increase its crew complement to the anticipated maximum of seven members, a larger escape vehicle is needed.

In response to the need for a modern escape vehicle capable of transporting all seven crew members, NASA is close to completing the development of a $4 billion escape vehicle called the X-38 that will replace the smaller Soyuz. The X-38, a twenty-nine-foot-long triangular pod, would use its body like a wing to glide back to Earth. Engineers have already tested two versions of the craft by dropping them from a B-52 aircraft over the California desert. The final version is anticipated for delivery in 2004. According to John Muratore, project manager for the X-38 project at Johnson Space Center in Houston:

> It's an all-electric vehicle that uses high-tech lasers and fiber optics to initiate many of the on-board sequences. Although it utilizes cutting-edge technology, it gets its roots from science we've already successfully used for many years. It's a great blend of both old and new knowledge.[15]

The architects of all space stations focused their time and energy on creating the best possible living environment in space. None of their ingenious inventions, from robotic arms to escape vehicles, would have value if the environment in which astronauts worked and lived were not optimized to allow for prolonged visits.

Chapter 3

Living in Outer Space

Space stations, especially the most recent ISS, were designed to keep the astronauts as comfortable as possible—the ISS modules are roomy, bright, and kept at a constant 70 degrees Fahrenheit. It is important that the crew members are comfortable because they are kept busy all their waking hours. In a typical day, crew members will spend twelve hours working, two exercising, two preparing and eating meals, and eight hours sleeping.

Despite the amenities provided, life in space reqiures considerable acclimation. Once on board a space station, the first order of business for novice astronauts is to become accustomed to the weightless environment, adjust to living in close quarters, and master new technologies necessary for carrying out routine daily activities. These three conditions, unique to all space stations, mean that the most basic and commonplace daily activities require rigorous attention, patience, and coordination.

Space Adaptation Syndrome

The first sensation experienced by three-quarters of all astronauts in weightlessness is space adaptation syndrome, more commonly known as space sickness. It is a form of motion sickness that occurs in spaceflights when astronauts are free to move about in the weightless environment. The syndrome did not occur on the lunar module or on early orbital flights because the astronauts were firmly strapped into small capsules.

Symptoms of space adaptation syndrome vary from one person to another but may include nausea, vomiting, anorexia, headache, malaise, drowsiness, lethargy, paleness, and sweating. The sickness is believed to be caused by sensory conflicts within and between the vestibular system—a collection of sensitive organs in the inner ear that maintain balance and orientation—and the visual system. Space physicians believe that when astronauts float and spin in a weightless environment, what their eyes see and what their vestibular organs sense lead to a neural mismatch that upsets the nervous system.

As intense as the symptoms can be, space sickness is usually of short duration, lasting from one to three days. Fortunately for astronauts, once they experience it, it never reoccurs. Unfortunately for them, however, although the sickness disappears relatively quickly, astronauts report that floating vomit is one of the least pleasant aspects of the first few days in space. When vomit is projected into the cabin, suction devices are deployed to capture and contain it.

Physicians have been able to reduce the occurrence of space sickness by employing countermeasures that include medications, head movement exercises to accelerate the process of adaptation, head restraints, and adjusting the vestibular system to weightlessness through biofeedback training.

Meals

Space sickness usually involves a dramatic loss of appetite, but eventually the astronaut begins eating again. When that happens, there are many challenges for the first-time visitor to the ISS. Meals on space stations present myriad problems because of weightlessness, a shortage of storage space, spoilage,

Café ISS

Ed Lu, the American commander on the ISS, enjoyed working on the space station as well as eating on it. While in space on the ISS, he wrote an article about food titled "Eating at Café ISS" for NASA's Web site. In the article, he describes some of the more interesting and amusing experiences of dining in a weightless environment.

We don't have a real kitchen up here, but we do have a kitchen table. You might wonder of what use a table is if you can't set anything down on it, but we have bungee straps and Velcro on the tabletop so you can keep your food containers, spoon, napkins, etc. from floating away. You can find Yuri [a Soviet cosmonaut] and I around the table 3 times a day. In fact the table, which is located in the Service Module, is kind of the social center of the ISS. Even though we only have 2 crew members now, it is where we congregate when we have time off. Of course there are no chairs around the table, what we do is float around the table while we prepare our meals and eat. There are a couple of handrails on the floor to slide your feet under to stabilize yourself.

As for utensils, the only utensil we use is a spoon. All of the food that requires a utensil to eat has some sort of sauce or at least some moisture to it, so it naturally sticks to the spoon. This is the same effect on the ground that allows drops of water to stick to windows, here it allows us to eat without having our food fly all over the place. This force isn't very strong, so you have to move fairly slowly when eating, or the food will literally fly right off your spoon and onto the wall.

The Russian drink packets are clear plastic and have a simple one-way valve where you add water; while the other side of the packets has a built in straw. The design is ingenious; you just cut off one end of the packet with scissors to open up the valve, slide the packet onto the water tap, turn on the water, mix well, and then use the scissors on the other end to open up the straw. The problem is that if you aren't careful, they have a tendency to leak and it is easy to get juice or tea all over yourself or the walls. The same property of liquids that lets them stick to your spoon also makes liquids stick to your face.

Much of the Russian-supplied food comes in cans. One of the advantages of cans is that if it is just for a short while, you can just let the can float as long as you are careful to keep an eye on where it is going. Remember that you don't have to worry about food spilling out of the can if it turns upside down!

and a shortage of water needed for food preparation. In spite of these obstacles to good food, astronauts look forward to meals more than any other daily routine.

Most foods are dehydrated to conserve space and are packed into compartments and freezers in the upper section and wardroom. Water, one of the most precious commodities, is particularly problematic since it cannot be compressed. It is also one of the heaviest commodities. As a result, equipment has been designed to capture and recycle water from exhaled air and even from urine.

All food is initially prepared, cooked, and packaged on Earth. Food is processed in a way that makes it stick to a spoon and keeps it from crumbling into hundreds of particles that could float away. Thick foods such as sauces, pastes, oils, peanut butter, and moist cake batter are used to bind dry, flaky foods together. Some foods are also selected for their natural ability to hold together. Tortillas, for example, are preferred to slices of bread because they create far fewer crumbs. Floating bread crumbs are more than a minor annoyance. On Skylab, astronauts complained about bread crumbs floating around the interior and getting stuck in filters or in their eyes.

To make food keep as long as possible, most of what is prepared for the ISS is freeze-dried, low-moisture, or thermostabilized—meaning that it has been heated to kill bacteria—and then sealed in airtight packages. To prepare them for consumption, foods require very little preparation. Even many beverages are packaged in a dehydrated form and then are slightly hydrated before they are consumed. The ISS's galley, like many kitchens on Earth, is equipped with water, microwave ovens, and refrigerators, allowing everyone on board to access more normal types of fresh food, including fruits, vegetables, and even ice cream.

Nutritionists ensure that the food astronauts eat provides them with a balanced supply of vitamins and minerals. Meals are eaten three times a day, but because astronauts expend less energy working in a weightless environment than they would on Earth, their caloric requirements are considerably lower than they would otherwise be. Calorie requirements differ from one astronaut to another. For instance, a small woman would require only about nineteen hundred calories a day whereas a large man would require about thirty-two hundred calories.

On board the ISS, more than one hundred food items are available to astronauts; half are provided by American nutritionists and half by their Russian counterparts. This is done to provide a fair mix of the foods from the two different cultures. American favorites are meatloaf and turkey with mashed potatoes and gravy, spaghetti, a variety of soups, brownies, peanut butter, and even ice cream. From the Russian chefs, favorites are a thick cabbage and beet soup called borscht; a selection of pickled meats; *baursaki*, which are small fried doughnuts made from unleavened dough; and *kazakh*, which are meat-flavored noodles.

There are many other types of foods an astronaut can choose from, such as fruits, nuts, chicken, beef, seafood, candy, and drinks, including coffee, tea, orange juice, fruit punch, and lemonade. Condiments are provided such as ketchup, mustard, and mayonnaise. Salt and pepper are available, but only in a liquid form because in orbit astronauts cannot sprinkle salt and pepper on their food; it would simply float away.

Some foods, such as brownies and fruit, can be eaten in their natural form. Other foods require adding water, such as macaroni and cheese or spaghetti. Some packaging, such as plastic tubes filled with mashed potatoes and gravy and soft ice cream, prevent food from flying away because they are sucked from holes at one end of the tubes. The food packaging is designed to be flexible, easier to

Living in Outer Space

NASA Food Categories

TYPE	DEFINITION	EXAMPLES
Rehydratable	Food is dehydrated to reduce weight and preserve it longer. These foods require added water and soaking before they can be eaten.	Soups, casseroles such as macaroni and cheese, appetizers such as shrimp cocktail, and breakfast items like scrambled eggs
Thermostabilized	Bacteria and enzymes in food are destroyed by heat processing. The foods are packaged in single servings and can be easily cut open after preheating.	Grilled chicken and ham, tomatoes and eggplants, and puddings
Intermediate Moisture	By limiting the amount of water they are packaged with, these foods last longer without spoiling but still retain a soft texture. They can be eaten immediately.	Dried beef and dried fruit (peaches, pears, and apricots)
Natural Form	Ready to eat and packaged in pouches. They do not require processing.	Nuts, granola bars, and cookies
Irradiated Meat	Cooked meat packaged in pouches and sterilized with ionizing radiation. These foods remain edible even when stored at room temperatures.	Beef steak and smoked turkey

Source: NASA Life Sciences Data Archive (http://lsda.jsc.nasa.gov).

use, and to maximize space when stowing or disposing of food containers.

Sleep

Just as calorie requirements are lower in orbit than on Earth, so too are sleep requirements less. When the time for sleep does come, weightlessness somewhat simplifies the process of bedding down. Since humans cannot sense an "up" or "down" in a weightless environment, they can sleep in any position. Since space is in short supply, designers of space station interiors can position astronauts to sleep vertically or horizontally. On the ISS, sleep compartments provide space for four people. The first person sleeps on the top bunk, the second on the lower bunk. A third person sleeps on the underside of the lower bunk, actually facing the floor. A fourth person sleeps vertically, attached to a wall with Velcro straps. Because the astronauts are in a weightless environment, mattresses are not needed. Instead, each bed consists of a padded board with a fireproof sleeping bag attached to it. Astronauts zip themselves inside the sleeping bags, generally leaving their arms out. Crew quarters also provide each astronaut with an individual light, communications station, fan, sound-suppression blanket, and sheets with weightlessness restraints for those who find the sleeping bags too warm. Pillows are available as well.

Sleep can be difficult to find from time to time. Much like on Earth, astronauts report waking up in

An astronaut writes at his ISS sleep station. Because of the weightless environment, ISS astronauts need only a padded board and sleeping bag to sleep comfortably.

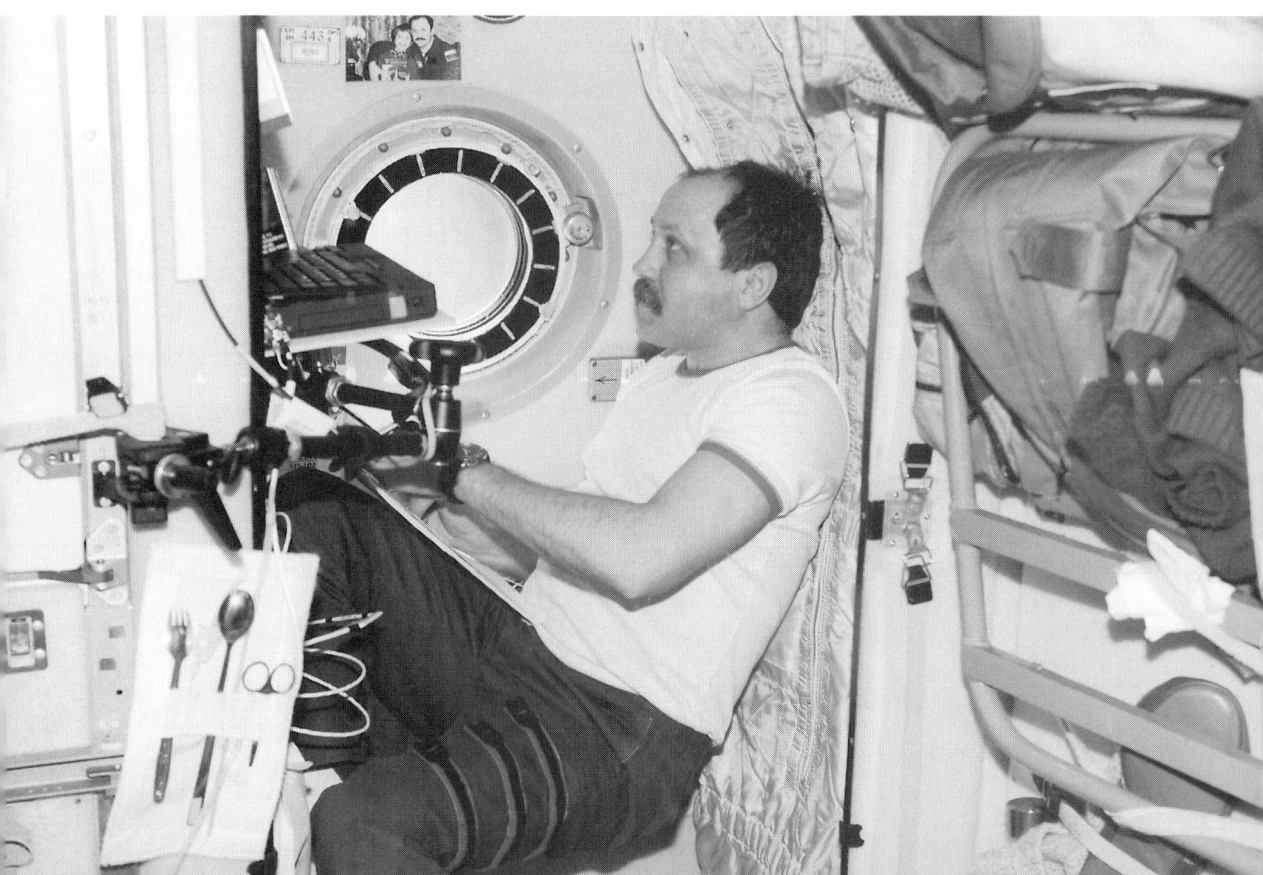

the middle of their sleep period to use the bathroom, and distractions can cause them to stay up late staring out the window. During their sleep period, astronauts have reported having dreams and nightmares. The close quarters can also result in sleep disruptions since crew members can easily hear each other; weightlessness does not, apparently, prevent snoring. In addition, sleeping near a window can be difficult since the Sun rises every ninety minutes as the station orbits Earth. The sunlight and warmth entering a window is enough to disturb a sleeper who is not wearing a sleep mask.

At no time are all crew members allowed to sleep at the same time; someone must always be awake to handle potential emergencies. Ground controllers actually decide when individual crew members go to bed. When it is time to wake up, the ground control sends wake-up music or a call to the crew. On America's Skylab, ground control picked a song for an astronaut each day. Sometimes a family member will request that controllers play a favorite song for their particular loved one on the ISS. In other cases, depending on the astronaut's own tastes, ground control may play rock and roll, country and western, or classical music. However, most of the time the wake-up call is unnecessary since most space station crew members use an alarm clock just as they might on Earth.

Exercise

No matter what sleep schedule a crew member adheres to, exercise is a critical part of the daily routine. Treadmills and ergometers, more commonly called stationary bikes, are used extensively by astronauts to maintain fitness. Such equipment has been used in space since Skylab in the 1970s, although they tend to cause a good deal of vibration. This can disrupt sensitive experiments elsewhere in the spacecraft, making sophisticated shock absorption systems necessary. Resistive exercise, a newer option to workouts, eliminates the vibration issue. Astronauts stress their bones and muscles by working against a resistive force, usually by pulling against strong bungee cords. Less motion is involved, and so there is very little vibration.

Strenuous exercise is vital for the heart. Over time, the astronaut's body responds to weightlessness by decreasing the amount of blood. Without regular strenuous exercise, an astronaut's heart will shrink, as it only has to pump this smaller volume of blood. This condition creates problems once an astronaut returns to Earth, when the heart must once again pull blood up to the brain against the force of gravity.

Personal Hygiene

Just as important to keeping fit is keeping clean. At the same time, sanitation is more difficult to maintain within the confines of a space station than on Earth. Studies conducted on early space station flights revealed that the populations of some microbes can increase rapidly under the combination of weightlessness and the confined spaces of an orbiting space station. The consequence is that many infectious diseases can easily spread to everyone aboard a space station. This is of particular concern since access to medical personnel is limited at best and pharmaceuticals are in short supply.

To prevent the growth of microorganisms, the eating utensils, dining area, toilet, and sleeping facilities in a space station are regularly cleaned. All potential sources of contamination must be carefully isolated. Since there is no washing machine aboard, soiled articles of clothing are sealed in plastic bags. Garbage and trash are also sealed in plastic bags, as are all food containers and used eating utensils, all of which is returned to Earth for laboratory analysis before it is either recycled or destroyed.

A Typical Day on the ISS

Astronauts on the ISS work six and a half days per week. Each day is carefully planned to optimize time for all needed activities, and the only relief from long days is a half day each Saturday.

Astronauts wake up at 7:00 Greenwich Mean Time, which is 2 A.M. in Houston and 11 A.M. in Moscow. Astronauts cannot rely on the usual daylight and nighttime cycles because in orbit and traveling seventeen thousand miles per hour, crew members experience sixteen sunrises and sunsets each day. Because of this rapid orbit rate, everyone must cover windows or wear masks to sleep.

After rising, the next order of business is using the bathroom and washing up. If all suction devices for the toilet and washbasin are working properly, astronauts finish in fifteen minutes and move to the breakfast area for the simplest and shortest of the day's three meals. If the suction plumbing is not working, someone may spend the remainder of the day fixing it, a common problem. At 8 A.M. sharp, the daily planning conference begins that involves a conference call with ground control centers in Houston and Moscow to review the plan for the day and answer any questions. Following that, everyone gets started on the day's work.

The work assignments vary week to week for each astronaut, but each will spend about twelve hours a day performing some experiment or part of an experiment. On the ISS, the most common work investigates how metal alloys crystallize as they cool in the weightless environment. Another major task involves work on plasma crystals, which are microscopic plastic spheres with electrical charges that repel each other and in the process form a regular lattice structure not found on Earth.

After work, or interspersed with experiments, each astronaut has certain housekeeping responsibilities scheduled. These are things like cleaning filters, performing periodic inspections of the emergency equipment, testing the water supply, and vacuuming out the air ducts.

Twice during the day—once in the morning and once in the late afternoon—each member completes a one-hour exercise program. After sweating on the treadmill or bike, they wash using towelettes impregnated with no-rinse soap or shampoo. At about 7 P.M. the second conference with Houston and Moscow is held to review the results of the day, after which dinner is eaten, the favorite activity of the day. Following dinner, each astronaut has a couple of hours of free time to send and read e-mails from home, take photographs out the window, listen to music, and write entries in personal journals. Finally, around 10 or 11 P.M., it is time for bed.

Each crew member has his or her own personal hygiene kit, which contains items such as a razor, shaving cream, hand cream, toothpaste, a toothbrush, a comb, nail clippers, deodorant, and other personal items, just as one might have on Earth. However, simple tasks like brushing teeth can be challenging in a weightless environment. The water that one would ordinarily use to wash out one's mouth cannot simply be spit out to drain away; any that escapes during brushing floats around in midair. To deal with this, astronauts use a freshwater hose followed by a vacuum hose to suction off the used water.

Although brushing teeth is a relatively simple task, other hygiene tasks prove time consuming and complicated under weightless conditions. Commonplace activities on Earth, such as shaving and hair cutting, for example, are slow, tedious processes because they must be done inside a plastic tent equipped with a vacuum device to suction away loose hair and whiskers. Escaped bits of hair are more than just unsightly. Any hair or whiskers floating about can lodge in sensitive electronic equipment, causing it to malfunction. Given the problems such activities present, most astronauts choose to avoid shaving and haircuts for as long as possible.

Some aspects of personal care, such as keeping clean, are not optional. In Skylab, astronauts actually showered using an enclosed shower stall. The stall was a cylinder with a collapsible fireproof canvas shower curtain for sides and a metal ring to secure it to the floor. When not in use, the whole assembly was collapsed and stored on the floor. To use the shower, astronauts would step inside the ring on the floor, raise the canvas curtain on a hoop, and attach it to the ceiling. Each astronaut was provided a ration of three gallons of water dispersed from a flexible hose with a push-button shower nozzle. The used water was contained within the stall and was vacuumed from the shower enclosure into a bag and then deposited in the waste tank. On the ISS, however, astronauts prefer quick sponge baths using washcloths or moistened towelettes.

Although the complexity of bathing sometimes acts as a deterrent to personal grooming, using the toilet is an even more complicated task—and one that cannot be avoided. The toilet on all space sta-

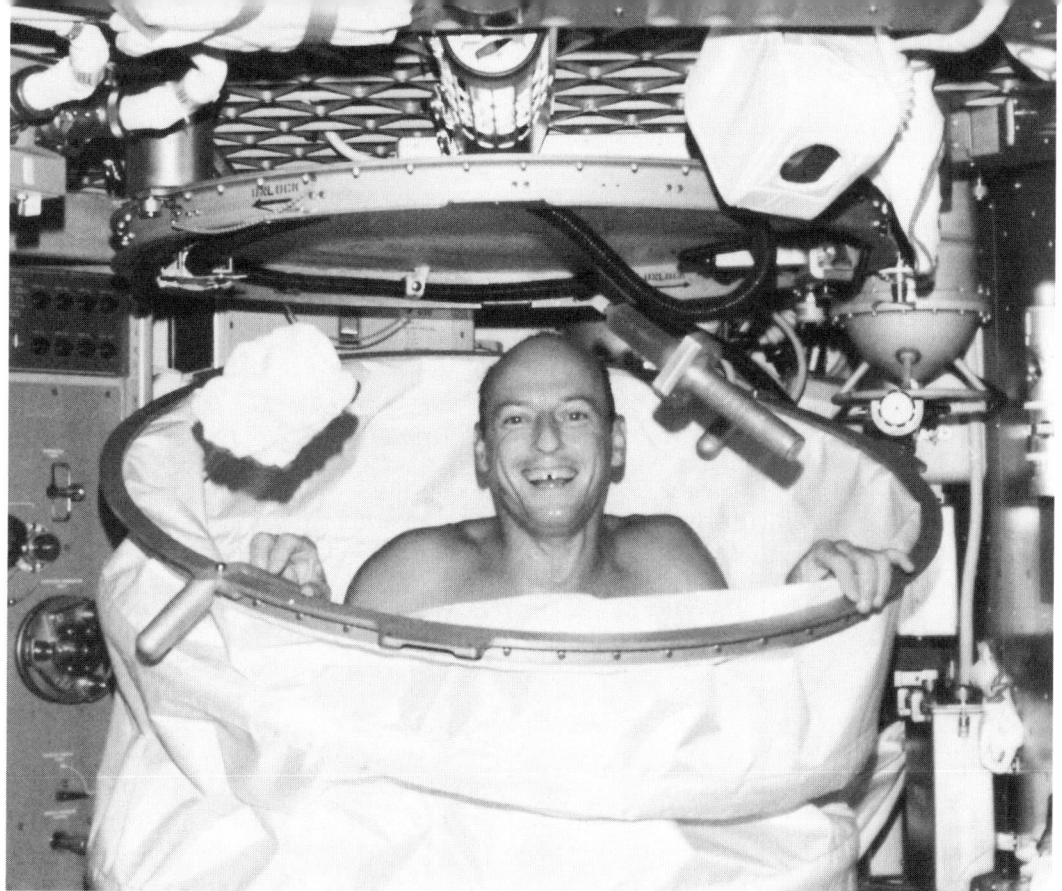

A Skylab astronaut smiles after a hot shower in the collapsible, sealed stall. Only three gallons of water per shower were allowed on the station.

tions is a small facility against a wall with only a partial door for privacy. The astronaut uses straps to keep from floating off the toilet seat; all urine and excrement are sucked into five-gallon plastic containers. A fan creates suction, doing the work that gravity does on the ground. The waste is returned to Earth, where it is analyzed as part of ongoing research into how the body functions in weightlessness.

Space Radiation

Not all hazards and inconveniences of space travel are as mundane and mechanical as using the bathroom and brushing one's teeth. In orbit, astronauts are exposed to radiation to a much greater extent than on Earth, where the atmosphere provides a shield for all living things. Of particular concern is the radiation emitted by the Sun, especially during periods when massive flares erupt from the Sun's surface. The radiation released during these massive

Space Suits

To explore and work in space, astronauts must take their environment with them because there is no atmospheric pressure and no oxygen to sustain life outside of their craft. Inside the spacecraft, the atmosphere can be controlled so that special clothing is not needed, but when outside, humans need the protection of a space suit.

The $12-million space suits used for space walks on Skylab and the International Space Station are a modular design so complex that users require an assistant to help put them on. The lower module, called the Lower Torso Assembly (LTA), roughly the equivalent of pants and boots, consists of a waist module, trousers module, and boots. The pieces are made of fabric but are joined together using metal bearing rings. The term *fabric* is really an understatement for the material, however. It contains a layer of urethane-coated nylon, followed by Dacron, neoprene-coated nylon, five layers of aluminized Mylar, and an outside layer of Teflon, Kevlar, and Nomex. Collectively, these many layers control internal temperature and protect the body against micrometeorite strikes that otherwise would easily penetrate the suit, causing a loss of pressure and oxygen, and pass through the astronaut's body, possibly causing death.

The counterpart to the LTA is the Hard Upper Torso (HUT), which is made of fiberglass and connects the arm module, glove module, and helmet module. The Primary Life Support System (PLSS) attaches to the back of the HUT. It resembles a backpack and provides the astronaut with oxygen and battery packs. The PLSS also controls the air pressure in the suit as well as the temperature of the oxygen and water that run through the garment to keep the astronaut cool. The HUT removes humidity, odors, and carbon dioxide from the air inside the suit and also carries the communication equipment and a multitude of sensors. A secondary oxygen pack attaches to the bottom of the PLSS for emergency oxygen and other life support functions. On the front of the HUT, astronauts carry a Display and Control Module, which keeps them informed about the status of the PLSS.

Apollo space helmets are formed from high-strength polycarbonate and Kevlar and are attached to the space suit by a pressure-sealing neck ring. Unlike earlier helmets, which were closely fitted and moved with the crew member's head, the Skylab helmet is fixed and the head is free to move within it. While walking in space, astronauts wear an outer visor assembly over the polycarbonate helmet to shield against eye-damaging ultraviolet radiation and to maintain head and face thermal comfort.

explosions passes through delicate human tissue and can damage cells.

The primary risk to astronauts comes in the form of an increased likelihood of cancer. Dr. Francis Cucinotta, director of Space Radiation Health at NASA's Johnson Space Center in Houston, says, "Younger women are particularly vulnerable to cell and tissue damage from space radiation. The greatest threat is an increased chance of developing breast, ovarian or uterine cancer."[16] Dr. Cucinotta adds that men's bodies overall are less sensitive to radiation, but even so, a forty-five-year-old male astronaut will only be allowed by NASA to spend a total of about 250 days in space. A clear understanding of the threat has not yet been achieved, however. According to physicians writing for the European Space Agency, "The long-term effects of space radiation on the human body . . . are still totally unknown."[17] This view is echoed by Dr. Paul Todd, chairman of the American Society for Gravitational and Space Biology, who adds, "This is not an easily solved scientific problem."[18]

Space Walks

Concerns over potential overexposure to radiation become even more acute when astronauts leave the shielded environment of their space stations and venture outside. For a variety of reasons, space walks, referred to by aerospace personnel as extravehicular activities (EVAs), are necessary. Such activities as attaching new modules, repairing equipment, and replacing worn-out parts all require some EVAs. ISS planners anticipate that the assembly of all modules will require about fourteen hundred hours of EVAs and about the same number over the lifetime of the ISS for a variety of repairs and adjustments.

Prior to departing the craft, astronauts don their pressurized space suits, which can sustain them for up to six hours at a time. Aboard the ISS, space suits have gloves with fingertip warmers for better dexterity, radios with multiple channels for communications, helmet-mounted floodlights and spotlights, internal controls for heat and cooling, and new multilayer fabrics to protect against extreme temperatures, ultraviolet radiation, and even micrometeorites. During an EVA, the weightless environment is a distinct advantage since each fully equipped space suit weighs 220 pounds on Earth.

Prior to departure, astronauts suit up in an airlock, which is a compartment that is sealed off from the space station. Then, all air is pumped out, which accomplishes two important objectives. First, it allows the astronauts to adjust to gas mixture differences in their space suits. Second, they gradually adjust to the change from the atmospheric pressure maintained in the space station to the dramatically lower pressure of space. After about forty minutes, the hatch is opened to the outside, the astronauts clip on a nylon cord that acts as a tether, and they begin their EVA. Against the possibility that the tether could become accidentally detached, NASA engineers developed a jet-powered backpack that allows free-floating crew members to fly back to the station.

EVAs are not undertaken without good reason, nor are they done without careful preparation. According to ISS astronaut Don Pettit, "Nothing happens fast. It takes several days to prepare for a space walk. Small details are important. We clean our visors and spread a thin layer of anti-fog on the inside surface. If there is too much anti-fog it can make your eyes sting and water; too little and it will fog up. It has to be just right if you want to see anything."[19]

Once in space, astronauts must move slowly using hand and toe holds welded to the exterior. Movement is slow because the bulky space suit contains hundreds of wires and cords that could snag or tangle. A space walk is no time to take chances, and when things go wrong, EVAs are sometimes aborted.

Living in Outer Space 63

Extravehicular Activity (EVA) Suits

Source: NASA (http://spaceflight.nasa.gov/station/eva/).

- **Helmet/Visor Assembly** (includes lamps, camera, and gold sun filter in visor)
- **Primary Life Support System** (backpack includes oxygen, power supply, fans, and radio)
- **In-Suit Drink Bag** (water-filled pouch with a straw, attached to the inside of the Hard Upper Torso)
- **Liquid Cooling and Ventilation Garment** (similar to long underwear)
- **Lower Torso Assembly** (includes pants, built-in boots, and rings for attaching safety tethers)
- **Communication Carrier Assembly** (fabric cap with built-in earphones and microphone)
- **Hard Upper Torso** (fiberglass shell beneath fabric)
- **Digital Displays and Control Module**
- **Gloves** (include heaters and loops for holding tools)
- **Maximum Absorption Garment** (adult-sized diaper with extra absorption material to collect urine)

The EVA suit has fourteen layers of nylon, Dacron, Mylar, neoprene, and fabric to protect astronauts from extreme heat, cold, and micrometeors.

Astronaut Jerry M. Linenger explains that each arm and leg movement during an EVA requires a great deal of thought and planning. There are plenty of hazards, and a single mistake could be catastrophic:

> A tear big enough to expose you to the full vacuum of space would be one of the most painful deaths imaginable. All the air would be sucked from your lungs. Blood would feel as if it was boiling in your veins, and your internal organs would go into seizure. A space walker must keep tethered to his spacecraft. There are no second chances.[20]

The thirty-year history of space stations has clearly established the viability of humans living and working in orbit. Dozens of astronauts living successfully in space for a total of thousands of days have laid a foundation for continued research on life in space. What at one time was a topic of conjecture has been conclusively and decisively answered by experimentation. During the 1970s physicians specializing in space medicine also set out to establish whether the human body could successfully function in a weightless environment. With that objective and others in mind, hundreds of medical experiments have revealed some interesting results.

Chapter 4

Space Medicine

Prior to the first flights of Salyut and Skylab, the effects of long-term exposure to a weightless environment were a matter of speculation. Aerospace engineers and space medicine teams from both the Soviet Union and the United States understood that unless humans could adequately adapt to weightlessness, hopes for more sophisticated space stations and long-duration spaceflights would never be realized. To this end, many of the experiments conducted aboard space stations involved determining, testing, measuring, and assessing changes in the conditions of the astronauts and cosmonauts themselves. In a very real sense, the researchers were the subjects of their own experiments. As one might expect, the early flights of Salyut and Skylab established the first thresholds for tolerance of weightlessness; Mir and the ISS have since tested those limits.

Blood and Fluid Distribution

The principal concern among space physicians regarding the functioning of the heart in weightlessness is the issue of blood and fluid distribution in the body. Physicians explain that under normal conditions, blood and other body fluids tend to pool in the legs. To counter this effect of gravity, veins in human legs have evolved valves that open and close to assist blood circulation back up to the heart. In orbit, however, blood pressure equalizes and fluids tend to reverse what they do on Earth and pool toward the head.

Music in Space

Space physicians are always aware of the importance of a calm environment on long space station deployments. They try to accommodate the psychological needs of the crew, especially during their personal time between dinner and bedtime. For those two to three hours, crew members are free to do whatever they wish. One of the favorite activities on the ISS, in addition to e-mailing family and friends, is playing musical instruments.

Astronaut Carl Walz once lived on the ISS for 196 days. Before he went up in 2001, Walz recalls in an interview with Karen Miller in her online article "Space Station Music," people asked him what kind of things he would be interested in taking along. Walz said, "Well, a keyboard would be nice. And they said, we'll look into that." He got his request.

A surprising number of astronauts play instruments. There was once even an astronaut rock-and-roll band. And a surprising variety of musical instruments have found their way into space; in addition to the keyboard, there has been a flute, a guitar, a saxophone, and even an Australian aboriginal wind instrument known as a didgeridoo. Astronaut Ellen Ochoa, a classical musician, brought her flute.

In Miller's article, Ochoa recalls, "When I played the flute in space, I had my feet in foot loops." In a weightless environment, even the small force of the air blowing out of the flute would be enough to move Ochoa around the shuttle cabin. In fact, even with her feet hooked into the loops, she could feel that force pushing her back and forth as she played. Still, she adds, "Music makes it seem less like a space ship, and more like a home."

Playing and listening to music is a favorite pastime for off-duty ISS astronauts.

This shift of blood and other fluids toward the head precipitates many problems. The brain interprets its increased blood supply as an increase in total fluid volume rather than simply a redistribution. In response to this misperception, the brain signals the kidneys and other organs to decrease the volume of blood and other fluids by pulling water out through increased urination. The decrease in fluid volumes is not in itself a problem, but the process in turn triggers losses of minerals such as calcium, which leads to loss of critical bone minerals.

Additional tests performed aboard both Skylab and the ISS indicate that an astronaut's blood volume decreases by 10 percent. Although it appears that fluid volume may stabilize at some reduced level, crew members must consume more water, a resource in short supply, to prevent dehydration.

One solution for maintaining normal blood and fluid distribution while in weightlessness is to wear pressure suits such as those worn during launch and reentry. American and Russian crews on the ISS have experimented with the regular periodic wearing of lower-body pressure suits in order to push fluids into the lower extremities. Although this has had some limited success in stabilizing blood volume, astronauts complain about the discomfort of the suits, which they say inhibits their work.

Cardiovascular Changes

In addition to the effects of weightlessness on fluid distribution, space physicians noticed changes to the cardiovascular system among the crews of all space stations. Initially detected during Skylab and Salyut missions, these changes included a lowering of the diastolic blood pressure—that is, the pressure during the heart's relaxation phase—and a tendency for fainting among space crews. On Salyut and Skylab, the volume of blood actually pumped by the heart was generally elevated during flight. Given the documented

10 percent drop in actual blood volume, this meant the heart was working harder than it did on Earth. This occurred in spite of a progressive decrease in cardiac size.

Other more precise measurements on the ISS using echocardiography confirmed these earlier findings and provided additional information. Echocardiography, in which ultrasound is used to make images of the heart chambers, valves, and surrounding structures, yielded remarkable results. Researchers discovered that the volume of the right ventricle, the chamber that pumps blood to the lungs, decreased by 35 percent during the first day of flight. Meanwhile, the left ventricle, the chamber that pumps blood to the rest of the body, increased in size by 20 percent during the first day, then decreased to 85 percent of its preflight volume during the second day. The volume of pumped blood varies dramatically, and the heart rate increases by 20 percent. As a result, space physicians realize that cardiac output increases substantially during the first day, then decreases to preflight level.

In addition to the use of echocardiography to evaluate the cardiovascular system, in-flight sampling of blood and urine affords researchers the ability to study the chemical and gas composition of the blood as well as the functioning of the kidneys, which filter waste from the blood. These tests reveal a decrease in the red blood cell count in returning astronauts, what is known as spaceflight induced anemia. Research also indicates changes in cellular morphology—that is, the shape of the cells. The normal shape of red blood cells is that of a disk slightly concave on both sides. This shape provides more surface area for the cell to absorb oxygen. When this shape changes to become slightly twisted or flattened, it causes a dramatic reduction in red blood cell absorption of oxygen as well as nutrients. Research also indicates that upon return to Earth, blood and fluid

Space Medicine

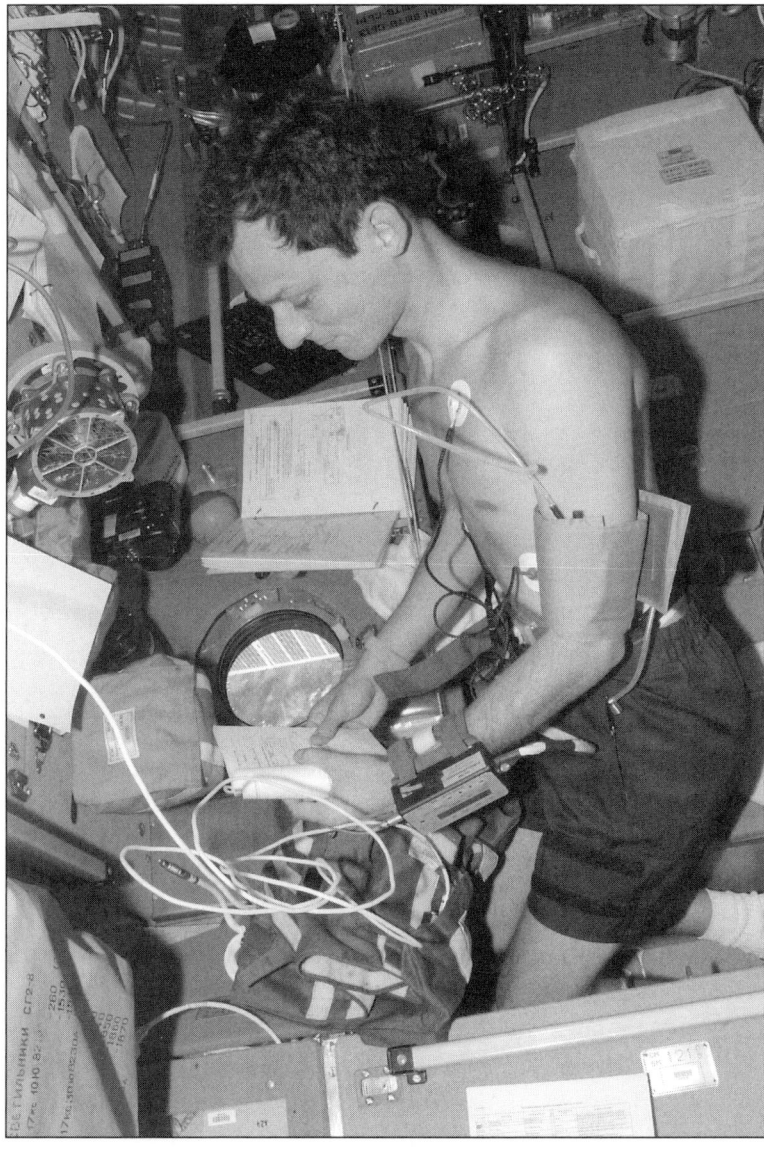

In space, an astronaut's heart undergoes changes in rhythm, output, and size. As a result, the heart must be monitored frequently.

levels return to normal, but cardiac output falls to subnormal levels. It takes several weeks for fluid volume, blood quality, cardiac size, and cardiac output to return to normal.

The Skeletal System
Just as prolonged weightlessness affects the cardiovascular system, so too does it affect the skeletal system.

Normally bone mass is deposited where it is needed and is reduced where it is not. Because the mechanical and pressure demands on bone are greatly reduced in a weightless environment, bone soon begins to dissolve and the resulting calcium, nitrogen, and phosphorus is absorbed and finally removed from the body by the kidneys.

Skylab astronauts lost an average of 8 percent of their bone mass in three months. Soviet cosmonauts, who usually remained in orbit for six months, averaged 15 percent loss, although one cosmonaut lost 20 percent while another lost only 8 percent. Such bone atrophy does not, however, affect the entire skeleton. Evidence on Skylab and the ISS suggests that non-weight-bearing bones such as the skull and fingers are not affected. In the legs and spine, however, which do bear weight on Earth, bone mass declines, as calcium is lost from both the cortical (outer) and trabecular (inner) bone tissue. Diminished bone mass becomes a problem when the astronaut returns to Earth. Also, since the blood carries excess calcium to the kidneys for elimination, the risk of kidney stones, which are made up of calcium, increases.

Space physicians attempt to control bone loss by requiring daily exercise. Aboard a space station, running on a treadmill offers the best workout possible for maintaining bone strength. The downside for astronauts is that an elasticized harness must be used to simulate gravity by pulling the user against the running surface. Astronauts find that such an arrangement is so uncomfortable that they are forced to take breaks every five or ten minutes.

Whether lost bone is regained once astronauts return to Earth's gravity is not entirely certain. Medical experts fear that the body's calcium balance might be restored before the bones have replaced all the lost minerals, resulting in permanent damage. Although cortical bone may regenerate, space physicians fear that loss of trabecular bone may be irreversible.

According to Dr. Jay Shapiro, team leader for bone studies at the National Space Biomedical Research Institute, "The magnitude of this effect has led NASA to consider bone loss an inherent risk of extended space flights."[21]

The experience of space travelers so far suggests that this risk is real. For example, when the Soviet cosmonaut Yuri Romanenko returned to Earth from Mir after completing his 326-day mission in 1987 (a record at that time), his bones were so brittle and weak that he had to be carried to a hospital. There, he was gradually allowed to increase weight on his skeletal structure over a period of several weeks because of fears that he might otherwise break many of

The International Space Station

To be built by:	United States, Russia, Japan, Europe, Canada, and Brazil
Construction:	Zarya module launched November 20, 1998; Unity module (Node 1) attached December 8, 1998; twelve additional modules, trusses, or other components have since been added, and construction is scheduled to continue through 2010
Weight:	206 tons* 500 tons (upon projected completion)
Habitable Volume:	15,000 cubic feet* 46,000 cubic feet (upon projected completion)
Size:	240 feet wide (span of solar arrays)* 146 feet long* 90 feet high* 356 feet wide (upon projected completion) 240 feet long (upon projected completion)
Crews:	The ISS has hosted twenty-two crew members (eight manned expeditions or crew exchanges)*
Missions:	Thirty-eight flights have been made to the ISS for crew exchanges or supplies (sixteen U.S. space shuttle visits, and twenty-two Russian flights)* One hundred and forty-eight science experiments have been completed or are in progress*

*As of spring 2004

Source: NASA (http://spaceflight.nasa.gov/station).

the bones in his lower extremities if he were allowed to walk too soon.

Muscles

In a weightless environment, muscles, like bones, atrophy from lack of use. Within the orbiting space stations, astronauts are able to move around by softly pushing against walls with a finger or toe and are able to move large loads without breaking a sweat. In 1982 Soviet cosmonauts returned from a 211-day mission on Salyut in obviously debilitated conditions. According to W. David Compton and Charles D. Benson in their book *Living and Working in Space: A History of Skylab*, "Although they had exercised daily, their muscles were so flabby that they were barely able to walk for a week, and for several weeks afterwards required intensive rehabilitation."[22]

Human muscle is of three types: smooth, cardiac, and skeletal. It is the effects of weightlessness on skeletal muscles, those that make movement of the whole body possible, that most concern space medicine specialists.

The bulk of skeletal muscles affected by gravity are located in the lower body. These are constantly under stress in order to keep the body upright. Other muscles also work against gravity—for example, those in the upper arms, shoulders, and back that are used for lifting and moving objects. These muscles, while used constantly on Earth, are hardly used in orbit, where even heavy objects float. When these muscles are not used, they atrophy. Muscle atrophy of 5 to 10 percent can occur by just eight days into a flight. Although muscle atrophy does eventually taper off over time, by the time astronauts have fully adapted to weightlessness, a large portion of muscle mass has been lost.

Experience has shown that all those returning to Earth following extended stays in space have difficulty standing or maintaining their posture. Astronauts also have coordination and walking problems until they are able to retrain their muscles to work against gravity. American astronaut John Blaha, who served on Mir, told fellow astronaut and author Jerry M. Linenger that when he returned from space, he had to be carried off the shuttle on a stretcher. Blaha went on to say that his muscles were so weak that "there was no way I could move. I felt like I weighed a thousand pounds. I could not even lift my arm, let alone stand up and walk. No way."[23]

The Psychological Effects of Space Life

Russian and American space physicians are just as concerned with the psychological effects of long-term stays on space stations as they are about the physical effects. Although the psychology of working in weightlessness is not a major concern, the psychological effects on space station crews of remaining in confined quarters for hundreds of days, far from friends and families, is a serious concern to NASA and other government agencies that deal with the ISS.

Stress has been a by-product of the isolation and close quarters common to all space stations. When psychological problems are discussed, the "twenty-four-hour mutiny" that occurred aboard Skylab is frequently brought up. For one twenty-four-hour period, astronaut Gerald Carr, Ed Gibson, and Bill Pogue refused to do any work, choosing instead to relax, look out the window, and rest. This unexpected rebellion by men acustomed to following orders is seen as evidence that long-duration spaceflights place great stress on astronauts, causing them to act in ways unimaginable on Earth.

Although all space station astronauts have experienced intense stress, some of the most noticeable forms of unsettling behavior have been seen on Mir. In 1996, for example, American astronaut John Blaha,

The Call of the Abyss

In the days of sailing ships, healthy young sailors would occasionally throw themselves from the boat and drown, overcome by a fascination with the deep, seemingly endless sea. This often-reported syndrome, labeled "the call of the abyss," seems to have a modern-day equivalent in spaceflight. Just as psychologists describe some people who are compelled to stand on the edge of precipices or tall bridges staring into the abyss and then jumping, more than one astronaut has expressed the same fascination by the free-falling view of space afforded by space walks.

Space walkers have expressed a strange sensation when floating in space with Earth below and the entire universe above them. Right from the start, some space walkers expressed a reluctance to return to the safety of their space station. America's first space walker, Ed White, had to be ordered back into his space station by the director of Mission Control. According to Dr. Tamarack R. Czarnik, who wrote an online article titled "Medical Emergencies in Space," White reportedly sighed and said to Mission Control, "It's the saddest moment of my life."

In 1977 this compulsion to stare with fascination into the void almost turned deadly for rookie cosmonaut Yuri Romanenko. During his stay aboard Salyut 6 with Georgi Grechko, a space walk was scheduled; Grechko would space walk while his partner, Romanenko would remain inside the airlock, monitoring medical readings. But Romanenko's curiosity got the better of him; he reportedly stuck his head out of the hatch and then began drifting farther and farther out. When he started drifting by, Grechko realized his friend's safety line was not attached, and Romanenko was drifting off into space. By leaning over as Romanenko drifted by, Grechko was able to grab hold of his loose safety line and pull him back in.

NASA is aware of this strange and interesting phenomenon. One of the reasons for the tether cord is to prevent space walkers from drifting off into space, where they would die within a few hours and then remain in orbit for years before falling back to Earth. Nonetheless, NASA officials remain vigilant about the possibility of an astronaut, mesmerized by the abyss of space, disconnecting his or her tether and drifting away.

Floating in the abyss of space is both exhilarating and exceedingly dangerous for astronauts.

After a few months onboard Mir, American astronaut John Blaha (left) began to exhibit hostility toward fellow crew members and other symptoms of serious depression.

who had been on board Mir for four months, began experiencing fits of anger, insomnia, and withdrawal from other crew members. According to fellow American astronaut Jerry M. Linenger, "He was hurting, he was, in essence, depressed."²⁴

Stress resulting when a fire broke out aboard Mir led Linenger himself to become increasingly withdrawn and isolated; eventually he even refused to participate in voice communications with ground control. Space historian Bryan Burrough observes in his book *Dragonfly: An Epic Adventure of Survival in Outer Space*, "Linenger's voice is high-pitched and shrill; he sounds as if he is on the verge of some kind of breakdown."²⁵

Many psychological problems on the ISS and Mir stemmed from cultural and political differences between the Russian and American crews. Part of the problem was the inability of the two crews to communicate effectively because no one was completely fluent in both Russian and English. This is of particular import for the ISS, where crew members of different nationalities must live together, perform experiments of various types together, and operate the spacecraft together in a confined place for three to six months. Political conflicts between Russian and American politicians over matters on Earth still occasionally spill over on the ISS, causing shouting matches among members of the crew.

Space station experiments since the 1970s have yielded solid results for understanding the psychological and physical stresses placed on astronauts. Some of what has been learned has also been applied to medicine on Earth for the benefit of the public. Although NASA managers and researchers are excited about their record to date, they are equally excited about a whole set of experiments in other disciplines as well.

Chapter 5

Research and Experiments

On each mission, when not subjecting themselves to medical and psychological testing, space station crews perform hundreds of scientific experiments. All experiments are selected by a panel of NASA scientists from thousands of suggestions. Each is then carefully planned and all needed hardware is assembled. Prior to liftoff, each crew rehearses the steps required for each experiment to minimize failures. Much is at stake: Multimillion-dollar projects can be rendered useless if an experiment is botched.

Scientists representing nearly all major branches of knowledge have jockeyed to gain permission to conduct experiments on Salyut, Skylab, Mir, and the ISS. Everyone has recognized that their unique environments, far beyond Earth's atmosphere and floating in weightlessness, hold extraordinary potential for new discoveries in many fields. Those fields given the highest priority have been astronomy, earth environmental study, material development, botany, combustion and fluid physics, and military reconnaissance.

The point of research in space, in the view of most scientists, is principally to improve human life on Earth. From this research they believe will come knowledge and discoveries that will change and improve everyone's lives on Earth, from the foods that people eat, the cars they drive, the computers they use, and even medical procedures used by physicians.

A Giant Leap for Astronomy

First with Salyut and Skylab, then with Mir, and today with the ISS, one of the key focuses of scientific exploration has been furthering human understanding of the cosmos. All space stations have carried instrumentation of various types on their missions miles above Earth to provide astronomers with clearer images of planets, stars, and galaxies than even the largest telescopes on Earth can offer.

The principal reason astronomers are interested in mounting their instruments on space stations is that they operate far above Earth's atmosphere, which obscures astronomers' views due to dust particles, changing temperatures, and moisture in the form of clouds, rain, and fog. In addition, light from large

A scientist on the ISS conducts research in one of the station's labs. The results of space research have helped improve the quality of life on Earth.

cities interferes by scattering throughout the dust and moisture droplets often found in the lower atmosphere. On space stations, however, far above Earth's murky atmosphere, many distant objects can be clearly seen and photographed.

One of the earliest attempts at placing a telescope on a space station occurred on Salyut for the purpose of investigating the Sun. Skylab followed with the Apollo telescope mount (ATM), a canister attached to the space station and containing a conventional telescope with lenses that could zoom in on a solar event such as sunspots. It also carried ultraviolet cameras that, thanks to sophisticated mounts, could be aimed steadily and precisely at any point on the Sun regardless of disturbances, such as those caused by crew movement. The instruments provided astronomers with thousands of remarkably detailed photographs of the Sun's surface and of solar flares.

With the launch of Mir, which carried state-of-the-art instrumentation, photographs deeper into space became possible. Soviet cosmonauts conducted a photographic survey of galaxies and star groups using the Glazar telescope. Because the telescope was pointed hundreds of millions of miles into deep space, far beyond the solar system, the amount of light being captured was so small that exposure times up to eight minutes were required to capture enough light for a single photograph. Under such circumstances, even the slightest vibrations from astronaut movements could shake the space station and ruin the photograph. As a result, all astronauts were required to sit, strapped into chairs, during these long exposures.

Of greatest excitement to astronomers today is a new generation of telescope, already built, tested, and secured on the ISS. This telescope, called the Submillimetron, is unique in three significant ways. First, as its name suggests, it detects and photographs very short wavelengths of light, much shorter than sunlight. These short microlight waves were emitted billions of years ago, when the universe was first formed. Astronomers believe that these images, then, are of cosmic bodies formed close to the beginning of the universe. Second, such a unique and precise instrument is designed to operate at supercold temperatures using liquid helium to chill sky-scanning equipment, thereby increasing the sensitivity of the Submillimetron's telescopic gear by slowing the motion of the molecules. A third unique feature allows for normal crew activity at all times, despite the extreme sensitivity of the equipment and extreme distances it photographs. The Submillimetron undocks from the ISS before it is used and then redocks for necessary maintenance. Astrophysicists interested in both the origin and ultimate fate of the universe are particularly interested in the Submillimetron's capabilities.

Investigating Environmental Hot Spots

Environmentalists and biologists recognize the value of space stations as a unique means to gain the broadest possible view of Earth as well as detailed views of particular environmental hot spots. When Earth is viewed from space through a variety of infrared and high-resolution cameras, natural resources can be identified, crops can be surveyed, and changes in the atmosphere and climate can be measured. Events on the surface, such as floods, oil spills, landslides, earthquakes, droughts, storms, forest fires, volcanic eruptions, and avalanches can be accurately located, measured, and monitored.

One of the earliest and most successful environmental projects carried out aboard a space station was the use of a scatterometer on Skylab. A scatterometer is a remote-sensing instrument capable of measuring wind speed and direction on Earth under all weather conditions. When it was activated on Skylab, the scatterometer captured wind speed and

Research and Experiments 81

direction data once a second and transmitted the data back to Earth. Engineers analyzed the data and used it to forecast weather, warn ships at sea of approaching heavy storms, assisted in oil spill cleanup efforts by accurately predicting the direction and speed the oil slick was taking, and notified manufacturers of hazardous chemicals of the safest times to ship their products.

Mir also proved its value to environmental science. One of Mir's modules, called "Priroda," a Russian word meaning "nature," was launched in April 1996. Priroda carried equipment to study the atmosphere and oceans, with an emphasis on pollution and other forms of human impact on Earth. It also was capable of conducting surveys to locate mineral resources and underground water reserves as well as studies of the effects of erosion on crops and forests.

To accomplish these ambitious objectives, environmental engineers loaded Priroda with active, passive, and infrared sensors for detecting and measuring natural resources. It carried several types of spectrometers used for measuring ozone and fluorocarbon (the chemical found in many aerosols) concentrations in the atmosphere. At the same time, equipment monitored the spread of industrial pollutants, mapped variations in water temperatures across oceans, and measured the height of ocean waves, vertical structure of clouds, and wind direction and speed.

When the ISS went into space in 1998, environmental studies were high on the list of projects for the astronauts to work on. From the ISS orbit, 85 percent of Earth's surface can be observed. Continuously monitoring and investigating Earth from space with an impressive array of high-tech instrumentation, the ISS has facilitated in the identification of many environmental problems. In 2001 the commander of the ISS, Frank Culbertson, shared with the British Broadcasting Corporation the many observations he and other astronauts had made after studying Earth's

The ISS Window

Designers of the ISS wished to add a special portal on one of the modules through which astronauts could gaze at and photograph Earth and neighboring planets. Gazing out into space was not new, but previous windows were made of glass that easily scratched, clouded, and discolored. In an effort to correct these defects, optical engineers created the Nadir window, named after the astronomical term describing the lowest point in the heavens directly below an observer.

Mounted in the U.S. laboratory module element of the space station, the twenty-inch diameter Nadir window provides a view of more than 75 percent of Earth's surface, containing 95 percent of the world's population. Designed by Dr. Karen Scott of the Aerospace Corporation, the high-tech five-inch-thick window is actually a composite of four laminated panes consisting of a thin exterior "debris" pane that protects it from micrometeorites, primary and secondary internal pressure panes, and an interior "scratch" pane to absorb accidental interior impacts. Each has different optical characteristics.

Scott headed a team of thirty optical engineers that used a five-hundred-thousand-dollar optical instrument to make fine calibration measurements on the window to ensure precise clarity free of distortion before installing it in the lab module. Tests conducted on the multiple layers of the window ensured that they would not distort under the varying pressure and temperatures common on the space station. After five days of extreme testing, the unique window was determined to have the characteristics that would allow it to support a wide variety of research applications, including such things as coral reef monitoring, the development of new remote-sensing instruments, and monitoring of Earth's upper atmosphere.

The ISS features a scratchproof portal through which the astronauts can gaze at Earth and other planets.

environment for four months. High above Earth, Culbertson made some startling observations:

> We see storms, we see droughts, we saw a dust storm a couple of days ago, in Turkey I think it was, and we have seen hurricanes. It is a cause for concern. Since my first flight in 1990 and this flight, I have seen changes in what comes out of some of the rivers, in land usage. We see areas of the world that are being burned to clear land, so we are losing lots of trees. There is smoke and dust in wider spread areas than we have seen before, particularly as areas like Africa dry up in certain regions.[26]

Cutting-Edge Cell Research

Since 2000, NASA has been conducting cellular research on board the ISS to take advantage of the weightless environment to study cell growth and the intricate and mysterious subcellular functions within cells. Traditionally, biologists study cells by slicing living tissue into sections of single-cell thickness. The drawback to this process, for as long as it has been practiced, is that the prepared specimens begin to die within a few hours as the cells begin to lose their ability to function normally. At best, researchers on Earth have only one day to scrutinize under microscopes the workings of minute structures within cells. The problem that occurs when single cells are removed from a living organ for examination is that microscopic structures crucial to the life of the cell collapse, causing the cell to cease functioning.

This research has primarily focused on the functioning of cells in the human liver, the organ that regulates most chemical levels in the blood and breaks down the nutrients into forms that are easier for the rest of the body to use. In a weightless environment slices of liver one-cell thick remain healthy and active for up to seven days, a significant advantage for researchers in space over those working on Earth. According to Dr. Fisk Johnson, a specialist in liver disease under contract with NASA, "Space is the gold-standard environment for this cutting-edge cell research. Only in space, a true microgravity environment, will we be able to isolate and study each of the individual factors impacting cell function."[27]

Once this advantage was discovered, the question then arose of how medical researchers on Earth could gain the same advantage. That question was answered by medical laboratories working with NASA that developed a device called a rotating bioreactor, which is capable of simulating a weightless environment on Earth. The rotating bioreactor works by gently spinning a fluid medium filled with cells. The spinning motion neutralizes most of gravity's effects, creating a near-weightless environment that allows single cells to function normally rather than collapse as they would otherwise do.

Utilizing the rotating bioreactor on Earth in the year 2002 scientists successfully accomplished long-term culturing of liver cells, which allows the cells to maintain normal functions for six days. One of the advantages of studying healthy cells for a long time is the ability to identify and match cellular characteristics to drugs that might cure particular diseases. According to Dr. Paul Silber, a liver specialist, "Our recent discoveries could lead to better, earlier drug-candidate screening, which would speed up drug development by pharmaceutical companies, and importantly, to a longer life for the 25,000 people every year waiting for a life-saving liver transplant."[28]

Creating Materials in a Weightless Environment

The weightless environment on space stations was of as much interest to materials scientists as to any others. Scientists are interested in a variety of physical properties of materials, such as melting points, molding characteristics, and the combining or separating of raw materials into useful products. Before the first space stations, materials scientists performed simple experiments of very short duration aboard plummeting airplanes and from tall drop towers. Through these studies, scientists discovered that gravity plays a role in introducing defects in crystals, in the combination of materials, and in other processing activities requiring the application of heat. Until the advent of space stations, however, they were incapable of sustaining a weightless environment long enough to thoroughly study these phenomena.

The advent of space stations allowed the study of new alloys, protein crystals for drag research, and silicon crystals for use in electronics and semiconductors. Materials scientists theorized that improvements in processing in weightlessness could lead to the development of valuable drugs; high-strength, temperature-resistant ceramics and alloys; and faster computer chips.

Using the ISS's Microgravity Science Glovebox, an astronaut studies the effects of weightlessness on various materials. In a weightless environment, scientists are able to remove impurities from most materials.

One of the Mir components, the Kristall module, was partially dedicated to experiments in materials processing. One objective was to use a sophisticated electrical furnace in a weightless environment for producing perfect crystals of gallium arsenide and zinc oxide to create absolutely pure computer chips capable of faster speeds and fewer errors. Although they failed to create absolutely pure chips, they were purer than those they could create within Earth's gravitational field.

More recently, fiber-optic cables are also being improved in weightlessness. Fiber-optic cables, vital for high-speed data transmission, microsurgery, certain lasers, optical power transmission, and fiber-optic gyroscopes, are made of a complex blend of zirconium, barium, lanthanum, aluminum, and sodium. When this blend is performed in a weightless environment, materials scientists are finding them to be more than one hundred times more efficient than fibers created on Earth.

In 2002 the ISS began the most complex studies of impurities in materials and ways to eliminate them in a microgravity environment. One of the more interesting causes of impurities, for example, is bubbles. On Earth, when metals are melted and blended, bubbles form. According to materials scientist Dr. Richard Grugel, "When bubbles are trapped in solid samples, they show up as internal cracks that diminish a material's strength and usefulness."[29] In a weightless situation, however, although bubbles still form, they move very slightly, and this reduces internal cracks. Secondarily, their slow movement allows researchers to study the effect of bubbles on alloys more easily and precisely.

According to Dr. Donald Gillies, NASA's leader for materials science, the studies of bubbles and other mysteries of materials production hold promise for new materials:

> We can thank advances in materials science for everything from cell phones to airplanes to com-

puters to the next space ship in the making. To improve materials needed in our high-tech economy and help industry create the hot new products of the future, NASA scientists are using low gravity to examine and understand the role processing plays in creating materials.[30]

New Discoveries

For centuries, physicists and chemists have been experimenting on a variety of elements and metals to discover new compounds and to improve existing alloys. They have also been aware that their experimental results are often affected by the containers they use and by the instruments that measure those results. Such contamination often invalidates experiments. Even worse, containers can sometimes dampen vibrations in a material or cool the sample too rapidly, throwing the validity of the experiment into doubt. In some cases, a metal is reactive enough to destroy its container, meaning that some materials simply cannot be studied on Earth.

When the first space stations went into orbit, physicists and chemists seized on the opportunity to conduct experiments within a weightless environment. If materials could be suspended in space during experiments, without the need for containers and eliminating the variables that the containers themselves imposed, far more accurate results would be allowable. Initial results of such experiments answered many questions that could not have been resolved on Earth. Of particular interest was the property of metals in a liquid state that causes them to resist solidifying, even at temperatures where they would be expected to do so. This phenomenon is called nucleation. According to Dr. Kenneth Kelton, a physics professor at Washington University in St. Louis, "Nucleation is the major way physical systems change from one phase to another. The better we

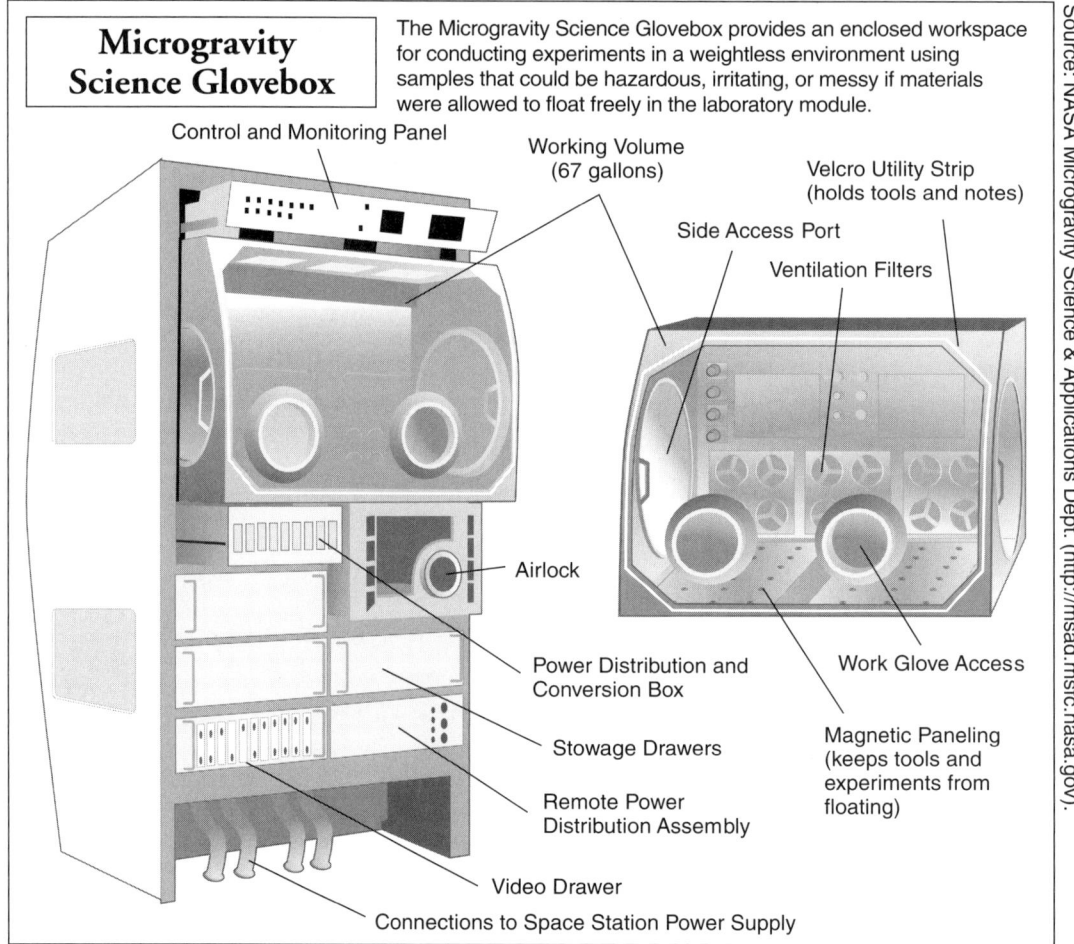

understand it, the better we can tailor the properties of materials to meet specific needs."[31]

Encouraged by the results of experiments carried out in space, engineers developed an apparatus on Earth that could duplicate a weightless environment for further research. NASA, joined by several private research companies, developed the electrostatic levitator (ESL), which is capable of suspending liquid metals without the sample touching the container and without the technicians handling equipment in ways that might alter results. Two practical applications using the ESL are the production of exceedingly smooth surfaces for computer and optical instrumentation and exceedingly pure metal for wires, making them capable of transmitting large volumes of data.

Greenhouses in Space

While materials scientists look to space station experiments in hopes of improving industrial processes on Earth, others are focused on investigating processes that might someday happen on a large scale in space. For example, botanists are studying the feasibility of crop cultivation on space stations in the belief that grains and vegetables may someday be needed in quantities large enough to supply deep space expeditions or even space colonies. To these ends, many experiments have been performed testing different gases, soils, nutrients, and seeds. One of them, called seed-to-seed cycling in a weightless environment, produced remarkably optimistic results. According to biologist Mary E. Musgrave:

> By giving space biologists a look at developmental events beyond the seedling stage, this experiment was an important contribution not only to gravitational biology, but also to the study of space life support systems. Data from this experiment on gas exchange, dry matter production and seed production provided essential information on providing a plant-based food supply for humans on long-duration space flights.[32]

Many of the botanical experiments in orbit have focused on the effects of weightlessness on plant growth and seed germination. Botanists had known for many years that seedlings on Earth display geotropism—that is, they respond to gravity by sending their roots down into the soil and stalks up above the ground. In addition, gravity affects the diffusion of gases given off by the plant, the drainage of water through soil, and the movement of water, nutrients, and other substances within the plant.

Early experiments aboard Skylab were not encouraging for those who hoped to grow plants in space. For example, researchers' speculations were confirmed that without gravity, the roots and stalks of plants could not correctly orient themselves. Some seedlings sent their roots above the soil and their stalks deep into the soil, with the result that they withered and died. And even those that did properly orient their roots and stalks often failed to produce seeds, a critical failure unanticipated by researchers.

In the mid-1980s, botanists performed an experiment to understand how seeds might survive weightlessness. Scientists sent 12.5 million tomato seeds into space and kept them there aboard Mir for four years. In 1990 the seeds were planted by botanists; many were also given to schoolchildren so they could make science projects of germinating them. Botanists discovered that a slightly higher percentage of seeds from space germinated than did seeds that had been kept on Earth and that almost all produced normal plants. These results were achieved even though the seeds had been exposed to radiation while in space.

A second significant experiment on the ISS sought to determine whether second-generation space plants would be as healthy as second-generation plants on Earth. Scientists analyzing the data concluded that the quality of second-generation seeds produced in orbit was lower than that of seeds produced on Earth, resulting in a smaller second-generation plant size. This diminished seed quality is believed to be caused by the different ripening mechanics inside the seed pod in weightlessness.

With so much evidence pointing to weightlessness as a hostile environment for plant production, botanists are a bit uncertain of the future of agriculture in space. One potential solution being investigated on the ISS is to grow plants without soil, a process known as hydroponics. In this process, the plants grow without soil, in a nutrient-rich solution.

A scientist studies plant growth in space. Research has shown that weightlessness is harmful to plant growth, throwing the viability of space agriculture into doubt.

Whether hydroponics can solve the problem of large-scale horticulture, though, is still uncertain.

Military Reconnaissance

In addition to their promise for scientists, space stations from the very beginning were seen as having military value. During the Cold War, when the United States and the Soviet Union jockeyed for political and military advantage on Earth, each country also looked to space stations to give them battlefield superiority. Although neither nation actually placed offensive weapons on board their space stations, both sought to exploit space stations' potential for reconnaissance.

All space stations have carried equipment capable of photographing objects 250 miles below. Photographs are detailed enough, for example, to allow analysts to determine the types and numbers of aircraft on aircraft carriers and to track troop movements on land.

Yet, military officials admit that so far, at least, space outposts can do little more than support more conventional military operations. At a meeting of the American Institute of Aeronautics held in Albuquerque, New Mexico, in August 2001, Colonel Steve Davis, an officer at Kirtland Air Force Base, said, "We're [the Air

Space Tourism

When NASA and the Russian Space Agency negotiated the initial agreement for the construction, deployment, and utilization of the ISS, no one gave consideration to using it as a tourist destination. From the inception of the project, all countries involved considered the ISS to be an orbiting laboratory dedicated to the study of a variety of scientific experiments and observations.

This somewhat parochial view was shaken in 2001 when the multimillionaire American businessman Dennis Tito expressed an interest in paying for a short vacation on the ISS to satisfy his own personal fascination with space. When NASA was notified of his interest and willingness to pay for a short visit to the spacecraft, his request was rejected on the grounds that the multibillion-dollar craft was for scientific purposes only. Recognizing that the Russians were short of money needed to continue their construction and launch costs, Tito approached them with an offer of $20 million.

Brushing aside NASA's objections, the Russians required Tito first to complete the standard training program before being blasted on what most called the most expensive vacation ever. In May 2001, when Tito docked at the ISS, several important milestones were achieved. These included the fact that a middle-aged civilian astronaut could easily survive space travel, that a space-tourism market did indeed exist, and that there was no longer a valid reason to discount the notion of space tourism.

Despite NASA's long-running opposition to his flight, which included preventing him from training with his Russian crewmates at the Johnson Space Center that triggered a minor international incident, Tito said he enjoyed his eight days in space and hoped that NASA would be more supportive in the future.

Millionaire civilian Dennis Tito's successful eight-day stay on the ISS introduced the possibility of a future tourism industry in space.

Force] still looking for that definitive mission in space; force enhancement is primarily what we're doing today." Davis added that there is increasing reliance on using space for military needs: "Space control is becoming more important as we have very high value assets in orbit. We depend on these assets and are interested in protecting them." Davis added that aboard one of the Soviet Union's early orbital piloted stations, it had a rapid-fire cannon installed. The military outpost was armed, Davis said, "so they could defend themselves from any hostile intercepts."[33]

Even the ISS is seen by some participating nations as having military value. An intergovernmental agreement on the ISS was first put in place in 1988, resulting in an exchange of letters between participating countries involved in the megaproject. Those letters state that each partner in the project determines what a "peaceful purpose" is for its own element. According to Marcia Smith, a space policy expert at the Congressional Research Service, a research arm of the U.S. Congress, "The 1988 U.S. letter clearly states that the United States has the right to use its elements . . . for national security purposes, as we define them."[34]

One of the more perceptive observations made when the first space stations flew into orbit was the potential that these floating laboratories might provide for investigating and solving a multitude of scientific questions. To a great degree, those making these observations were correct. Nearly every branch of science jumped on the space station bandwagon with proposals to investigate a host of questions. As the twenty-first century pushes forward, many problems of living in space have been solved while others remain elusive. The question being asked more frequently than ever is whether the costs of the many space stations and their experiments have returned enough benefits to taxpayers to continue the space station program.

Epilogue

Have Space Stations Met Expectations?

Although space stations have been functioning since the 1970s, during which time thousands of experiments have been performed, many observers question the value of this research. This skepticism is voiced by citizen groups questioning whether the research done has civilian application, by legislators questioning whether the billions of dollars spent might not be better spent elsewhere, and by scientists who believe the huge expenditures for a relatively small number of big projects might be better spent on a larger number of small projects.

When asked what good has come from space stations, managers of the U.S. space program often find it difficult to answer by citing tangible results. Bob Marshall, director at the Marshall Space Flight Center (MSFC) for NASA, found himself in this dilemma when he gave this rather perplexing response: "The main reason we're building the International Space Station is not because of what I can tell you we're going to do with it, which I can't. The main reason is because I can't tell you what we're going to do with it. And if you don't ever do it, you'll never find out."[35] In keeping with this statement that no one at NASA had a clearly defined objective for the ISS was a

Have Space Stations Met Expectations?

similar comment made about Skylab by space historians W. David Compton and Charles D. Benson. They assert in their book *Living and Working in Space: A History of Skylab* that when Skylab was designed and built, they did not have a clear idea what to do with it: "The Center [MSFC] was, as the official Skylab history has suggested, a tremendous solution looking for a problem."[36]

Cost overruns have added to concerns regarding the usefulness of the ISS. Congress has allocated many

Although critics of the ISS condemn the station's exorbitant operating expenses, its supporters believe that the station's research potential justifies its high cost.

tens of billions of dollars for space stations. In 1995, when NASA revealed its budget for the ISS to be $8 billion, U.S. representative Tim Roemer revealed that, according to independent government estimates, the true cost would be $72 billion. Roemer, along with many of his colleagues and constituents, believed that the money might be better spent:

> We're now looking at a start-to-finish cost of $72 billion—$72 billion, and I think our constituents would be very interested in hearing as we contemplate some very difficult cuts in our budget, whether we're looking at Medicare cuts, whether we're looking at cutting drug-free schools, whether we're looking at cutting farm programs.[37]

Opposition to the ISS has also come from the scientific community. Many scientists are uncertain as to whether such an expensive project is necessary. Much research, they argue, could have been done on Earth or remotely, using much cheaper satellites. Dr. Robert Park, a physics professor at the University of Maryland and the director of public information for the American Physical Society, has taken a strong stand against the space station. He argues that, "Its few scientific objectives have, for the most part, already been attained, and that the research planned for the station could be conducted aboard unmanned platforms and the Shuttle for far less money."[38]

Lack of practical applications for space station experiments is another common complaint. Many had hoped that there might be hundreds of better and cheaper goods for consumers because of space station research, yet so far, skeptics contend that little has trickled down to consumers. The cost of solar panels for consumers, for example, remains prohibitively high even though they have been used on all space stations. Besides, critics point out, NASA originally advertised that private companies would be able to perform experiments on the ISS for about four hundred dollars per pound of apparatus, yet the actual costs are closer to ten thousand dollars per pound, far too expensive for the research and development budgets of most companies. According to Senator John McCain in August 2003, "There is no doubt that the enthusiasm for the whole space effort has waned over the years. Most Americans don't know what we are doing in space."[39]

Despite such widespread skepticism, supporters of NASA point out that the ISS is a collaboration of many partners working together to create a world-class, state-of-the-art orbiting research facility. The station, they say, will afford scientists, engineers, and entrepreneurs an unprecedented platform on which to perform complex and long-duration experiments in the unique environment of space. And, they add, the ISS is much more than a world-class laboratory; it is an international human experiment—an exciting city in space—a place where much can be learned about how to live and work "off planet."

Thus far, the debate continues, as does the research on the ISS.

Notes

Introduction: The Road to Space Stations
1. Quoted in Mike Wright, "The Disney–Von Braun Collaboration and Its Influence on Space Exploration," History Office, Marshall Space Flight Center, 2003. www.history.msfc.nasa.gov.
2. Quoted in Richard Stenger, "Man on the Moon: Kennedy Speech Ignited the Dream," CNN.com, May 25, 2001. www.cnn.com.
3. Gary Lofgren, correspondence with author, September 4, 2003.
4. Quoted in Giovanni Caprara, *Living in Space: From Science Fiction to the International Space Station.* Buffalo, NY: Firefly Books, 2000, p. 32.
5. Andrew Dunar and Stephen Waring, *Power to Explore: A History of the Marshall Space Flight Center, 1960–1990.* Washington, DC: U.S. Government Printing Office, 1999, p. 137.

Chapter 1: Modest Beginnings: Salyut and Skylab
6. Quoted in Dunar and Waring, *Power to Explore*, p. 206.

Chapter 2: A Quantum Leap in Technology
7. Quoted in Dunar and Waring, *Power to Explore*, p. 115.
8. David M. Harland, *The Mir Space Station: A Precursor to Space Colonization.* New York: John Wiley & Sons, 1997, p. 60.
9. Harland, *The Mir Space Station*, p. xvii.
10. Quoted in BBC News, "The World's Future in Space," October 20, 2000. www.news.bbc.co.uk.
11. Caprara, *Living in Space*, p. 147.
12. Quoted in Tony Phillips, "Looking Forward to the ISS," NASA, April 2000. www.science.nasa.gov.
13. Al Feinberg, correspondence with author, September 16, 2003.

Notes

14. Buzz Aldrin, "America's Space Program: What We Should Do Next," *Popular Mechanics*, May 1, 2003, p. 110.
15. Quoted in NASAexplores, "The First-Ever Space Rescue Vehicle," March 2001. www.nasaexplores.com.

Chapter 3: Living in Outer Space

16. Quoted in Irene Brown, "Space Radiation Becomes Burning Issue," Discovery Channel, September 2003. www.dsc.discovery.com.
17. Quoted in European Space Agency, "Life Sciences: Research Announcement for the Utilization of the International Space Station," 2002. www.estec.esa.nl.
18. Quoted in Brown, "Space Radiation Becomes Burning Issue."
19. Quoted in Tony Phillips, "Getting Ready for a Space Walk," NASA, January 2003. www.science.nasa.gov.
20. Quoted in Damon Wright, "Danger in Space," FirstScience.com, January 2003. www.firstscience.com.

Chapter 4: Space Medicine

21. Quoted in Doug Hullander and Patrick L. Barry, "Space Bones," FirstScience.com, 2003. www.firstscience.com.
22. W. David Compton and Charles D. Benson, *Living and Working in Space: A History of Skylab*. Washington, DC: National Aeronautics and Space Administration Special Publication 4208, 1983, p. 324.
23. Quoted in Jerry M. Linenger, *Off the Planet: Surviving Five Perilous Months Aboard the Space Station Mir*. New York: McGraw-Hill, 2000, p. 222.
24. Linenger, *Off the Planet*, p. 126.
25. Bryan Burrough, *Dragonfly: An Epic Adventure of Survival in Outer Space*. New York: HarperCollins, 1998, p. 448.

Chapter 5: Research and Experiments

26. Quoted in BBC News, "Astronaut Sees Earth Changes," August 31, 2001. www.news.bbc.co.uk.
27. Quoted in ScienceDaily, "International Space Station Research to Study Treatments for Liver Ailments," 2002. www.sciencedaily.com.

28. Quoted in ScienceDaily, "International Space Station Research to Study Treatments for Liver Ailments."
29. Quoted in SpaceDaily, "ISS Ready to Crank Out the Material," June 2002. www.spacedaily.com.
30. Quoted in SpaceDaily, "ISS Ready to Crank Out the Material."
31. Quoted in Marshall Space Flight Center, "NASA Experiments Validate 50-Year-Old Hypothesis," June 2003. www1.msfc.nasa.gov.
32. Quoted in Mary E. Musgrave, "Developmental Analysis of Seeds Grown on Mir," National Aeronautics and Space Administration, August 1999. www.spaceflight.nasa.gov.
33. Quoted in Leonard David, "New ISS Duty: A Military Outpost?" Space.com, September 2001. www.space.com.
34. Quoted in David, "New ISS Duty."

Epilogue: Have Space Stations Met Expectations?
35. Quoted in Dunar and Waring, *Power to Explore*, pp. 527–28.
36. Compton and Benson, *Living and Working in Space*, p. 5.
37. Quoted in Federation of American Scientists, "International Space Station Authorization Act of 1995," July 28, 1995. www.fas.org.
38. Quoted in Jenna Minicucci, "Review of NASA's International Space Station," American Geological Institute, August 13, 1997. www.agiweb.org.
39. Quoted in Sheryl Gay Stolberg, "Shuttle Report Spurs a Debate in Congress," *New York Times*, August 28, 2003, p. B-23.

For Further Reading

Peter Bond, *Zero G: Life and Survival in Space.* New York: Cassell, 1999. This book provides vivid insights into the experiences of traveling in a rocket and then long-term life on space stations. Bond accentuates the trials, dangers, excitement, and joys of travel and life in space. The book includes many color photographs of both Russian and American space stations.

Bryan Burrough, *Dragonfly: An Epic Adventure of Survival in Outer Space.* New York: HarperCollins, 1998. This book is a retelling of what transpired when American astronauts joined the Russians on Mir as well as their background, training, and personalities. The book reveals the extent to which the Americans were not prepared to understand the workings of Mir nor the culture of Russian cosmonauts.

Mary M. Connors, Albert A. Harrison, and Faren R. Akins, *Living Aloft: Human Requirements for Extended Spaceflight.* Washington, DC: NASA Scientific and Technical Information Branch, Special Publication 483, 1985. This book covers perhaps some of the most interesting, but least-documented, aspects of spaceflight. The focus on biomedical and personal experiences of astronauts makes for useful and enjoyable reading.

Collin Foale, *Waystation to the Stars.* London: Headline Books, 2000. This is the story of British astronaut Michael Foale aboard Mir, as related by his father, Collin. It is an interesting book based on Michael's hundreds of e-mails to his parents and his personal journal of life aboard Mir. The book accurately recounts the Mir electrical fire, collision, and multiple computer failures. In spite of the tense moments, the book also documents the successes and friendships with American and Russian crew members.

Roger D. Launius, *Space Station: Base Camps to the Stars.* Washington, DC: Smithsonian Institution, 2003. In this well-illustrated book, Launius, a former NASA chief historian, details the developments of space stations from their origins to the ISS. He describes the public relations efforts of Wernher von Braun to promote space colonization, describes the science and politics of the earlier space stations Skylab and Salyut, and speculates on the future of the ISS and space exploration in general.

David J. Shayler, *Skylab: America's Space Station.* Chichester, England: Springer-Praxis, 2001. In this book, the author chronicles the evolution of Skylab; its infrastructure on the ground, including astronaut training; each of the three manned missions; a summary of the results of each mission; and the successes and lessons learned from failures.

Works Consulted

Books

Peter Bond, *The Continuing Story of the International Space Station.* Chichester, England: Springer-Praxis, 2002. Bond's book describes the development and evolution of space stations, with particular emphasis on the International Space Station. He begins his book with the revolution that began in 1970 when Salyut 1, the world's first space station, was sent into orbit and ends it with speculation about the future of space exploration.

Giovanni Caprara, *Living in Space: From Science Fiction to the International Space Station.* Buffalo, NY: Firefly Books, 2000. This book covers the history of space habitation from early theoretical speculation to the first launchings of modules for the International Space Station. The strength of the book is its ample photographs that complement the historically accurate text.

W. David Compton and Charles D. Benson, *Living and Working in Space: A History of Skylab.* Washington, DC: National Aeronautics and Space Administration Special Publication 4208, 1983. This book is one of the most comprehensive books that focuses on the politics and economics of NASA during the Skylab era. It is a well-documented and thorough view into the inner workings of NASA.

Andrew Dunar and Stephen Waring, *Power to Explore: A History of the Marshall Space Flight Center, 1960–1990.* Washington, DC: U.S. Government Printing Office, 1999. This book provides an in-depth account of the Marshall Space Flight Center and the engineers responsible for much of America's space program for thirty years. The book is extremely well documented, revealing that politics and economics were stronger motivators for exploring space than scientific discovery.

David M. Harland, *The Mir Space Station: A Precursor to Space Colonization.* New York: John Wiley & Sons, 1997. This book explores the development and operation of Mir, focusing on the engineering technology aspects of constructing and operating an orbital complex designed to be occupied by humans for long periods of time. This book also includes an excellent selection of photographs.

David M. Harland and John E. Catchpole, *Creating the International Space Station.* Chichester, England: Springer-Praxis, 2002. This work presents a comprehensive historical background, rationale for, and events leading to the construction of the ISS. The authors describe the orbital assembly of the ISS and experiments performed in the various laboratory modules. They also provide an account of the long-term stresses and strains of building the ISS on the U.S.-Russian relationship.

Jerry M. Linenger, *Off the Planet: Surviving Five Perilous Months Aboard the Space Station Mir.* New York: McGraw-Hill, 2000. Linenger is an American astronaut who spent five months aboard the Russian space station Mir. Linenger's account of that experience offers a balanced view of life aboard an aging space station. It is full of fascinating details about everyday life in a confined craft and how people survive psychologically and relate to one another under prolonged isolation and confinement. Linenger also provides insights into the bizarre relationship between the crew and mission control that may have kept him and his Russian comrades in constant danger.

Periodicals

Buzz Aldrin, "America's Space Program: What We Should Do Next," *Popular Mechanics*, May 1, 2003.

Sandy Fritz, "Beyond MIR," *Popular Science*, August 1, 1994.

Sheryl Gay Stolberg, "Shuttle Report Spurs a Debate in Congress," *New York Times,* August 28, 2003.

Works Consulted

Internet Sources

Jim Banke, "Space Age Materials Protect NFL Players from Harm," Space.com, September 2003. www.space.com.

Patrick L. Barry, "Voyage of the Nano-Surgeons," NASA, January 2002. www.science.nasa.gov.

BBC News, "Astronaut Sees Earth Changes," August 31, 2001. www.news.bbc.co.uk.

———, "The World's Future in Space," October 20, 2000. www.news.bbc.co.uk.

Irene Brown, "Space Radiation Becomes Burning Issue," Discovery Channel, September 2003. www.dsc.discovery.com.

Tamarack R. Czarnik, "Medical Emergencies in Space," Mars Society, 2002. www.mars.complete-isp.com.

Leonard David, "New ISS Duty: A Military Outpost?" Space.com, September 2001. www.space.com.

European Space Agency. "Life Sciences: Research Announcement for the Utilization of the International Space Station," 2002. www.estec.esa.nl.

Federation of American Scientists, "International Space Station Authorization Act of 1995," July 28, 1995. www.fas.org.

Doug Hullander and Patrick L. Barry, "Space Bones," FirstScience.com, 2003. www.firstscience.com.

Ed Lu, "Eating at Café ISS," NASA, July 2002. www.spaceflight.nasa.gov.

Marshall Space Flight Center, "NASA Experiments Validate 50-Year-Old Hypothesis," June 2003. www1.msfc.nasa.gov.

Karen Miller, "Space Station Music," NASA, 2002. www.science.nasa.gov.

Jenna Minicucci, "Review of NASA's International Space Station," American Geological Institute, August 13, 1997. www.agiweb.org.

Mary E. Musgrave, "Developmental Analysis of Seeds Grown on Mir," National Aeronautics and Space Administration, August 1999. www.spaceflight.nasa.gov.

NASAexplores, "The First-Ever Space Rescue Vehicle," March 2001. www.nasaexplores.com.

Tony Phillips, "Getting Ready for a Space Walk," NASA, January 2003. www.science.nasa.gov.

———, "Looking Forward to the ISS," NASA, April 2000. www.science.nasa.gov.

ScienceDaily, "International Space Station Research to Study Treatments for Liver Ailments," 2002. www.sciencedaily.com.

Melissa Snowden, "Russian Space Dogs," Silverdalen, 2003. www.silverdalen.se.

SpaceDaily, "ISS Ready to Crank Out the Material," June 2002. www.spacedaily.com.

Richard Stenger, "Man on the Moon: Kennedy Speech Ignited the Dream," CNN.com, May 25, 2001. www.cnn.com.

Damon Wright, "Danger in Space," FirstScience.com, January 2003. www.firstscience.com.

Mike Wright, "The Disney–Von Braun Collaboration and Its Influence on Space Exploration," History Office, Marshall Space Flight Center, 2003. www.history.msfc.nasa.gov.

Index

airlock, 17, 62, 74
Aldrin, Buzz, 47
alloys, 85
aluminum, 86
animals, 11
antennae, 17
Apollo, space helmets used on, 60
Apollo telescope mount (ATM), 14, 79
architecture, 16–17
Armstrong, Neil A., 12
astronauts
   Armstrong walks on the Moon, 12
   communication between ground control and, 12
   see also International Space Station
astronomy, 78–80
Atlantic Monthly (magazine), 9
atrophy, 72–73

backpack, 62
barium, 86
baursaki, 52
bed, 54
Benson, Charles D., 72, 95
bioreactor, 84
Blaha, John, 73, 75
blood, 65, 67–68
bone mass, 70
borscht, 52
brain, 67
bread crumbs, 51
"Brick Moon, The" (Hale), 9
British Broadcasting Corporation (BBC), 81, 83
bubbles, 86

Burrough, Bryan, 75

cables, 86
calcium, 67, 70
call of the abyss, 74
calories, 52
cameras, 13
cancer, 61
cannon, 93
cans, 50
Caprara, Giovanni, 40
carbon dioxide, 20
Carr, Gerald, 73
cell research, 83–84
cellular morphology, 68
ceramics, 85
chairs, 50
Clinton, Bill, 36
Cold War, 92
Collier's (magazine), 9
Columbus Orbital Facility, 41
communication, 12
Compton, W. David, 72, 95
computer chips, 85–86
computers, 23
contamination, 87
Core module, 32–34
crystals, 85–86
Cucinotta, Francis, 61
Culbertson, Frank, 81, 83
Czarnik, Tamarack R., 74

Dacron, 60
Davis, Steve, 91, 93
Dextre, 46
docking, 23–24
doctors, treatment of space sickness by, 49

dog, 11
Dragonfly: An Epic Adventure of Survival in Outer Space (Burrough), 75
dreams, 55
drinks, 20, 50
drugs, 85
Dunar, Andrew, 14
dust, 83

Eagle (lunar module), 12
Earth, 82–83
"Eating at Café ISS" (Lu), 50
echocardiography, 68
Eisenhower, Dwight D., 11
electricity, 22
electrostatic levitator (ESL), 88–89
emergencies, 42, 46–47, 55
environmental hot spots, 80–81, 83
ergometers, 55
escape vehicle, 41–42, 46–47
European Space Agency (ESA), 40–41, 61
exercise, 55–57, 70
Exposed Facility, 41
extravehicular activities (EVAs), 17, 61–64

fabric, 60
Feinberg, Al, 45
fiber-optic cables, 86
fire, 24, 75
fluid distribution, 65, 67
food, 12, 35–36, 49–53
Functional Cargo Block (FGB), 41

Gagarin, Yury, 11
gallium arsenide, 86
garbage, 21, 56
Gibson, Ed, 73
Gillies, Donald, 86–87
Glazar telescope, 79
glovebox, 85, 88
Goldin, Daniel, 38–39
gravity, 89
Grechko, Georgi, 74

greenhouses, 89–91
ground control, 12, 55
Grugel, Richard, 86
guns, 13
gyroscopes, 86

hair, cutting of, 58
Hale, Edward Everett, 9
handrails, 50
Hard Upper Torso (HUT), 60, 63
Harland, David M., 35–36
hatch, 20, 62, 74
heart, 56, 67–69
helmets, 60
High Energy Astronomy Observations (HEAO), 14
hurricanes, 83
hydroponics, 90–91
hygiene, 56–59

International Space Station (ISS)
  building of
    addition of escape vehicle, 46–47
    countries' contributions for, 40–42
    leading-edge technologies on, 42–45
    proposal for, 36, 38–40
    use of robotic arm in, 45–46
  life on
    exercise, 55–57
    meals, 49–53
    personal hygiene, 56–59
    playing musical instruments, 66
    sleeping, 53–55
    space adaptation syndrome, 48–49
    space radiation, 59, 61
    space suits, 60, 63
    space walks, 61–64
    typical workday, 57
  psychological problems on, 75–76
  windows on, 82
  see also research

Index

Japanese Experiment Module (JEM), 41
Johnson, Fisk, 84
Johnson Space Center, 92

kazakh, 52
Kelton, Kenneth, 87–88
Kennedy, John F., 12
Kevlar, 60
Khrushchev, Nikita, 11
Kibo, 41
kidneys, 67, 68
Kingsbury, James, 27–28
Kristall module, 86
Kvant module, 33

Laika, 11
lanthanum, 86
Large Space Telescope, 14
lasers, 86
left ventricle, 68
Liebermann, Randy, 9
Linenger, Jerry M., 64, 73, 75
liquids, 21
liver, 83–84
Living and Working in Space: A History of Skylab (Compton and Benson), 72, 95
Living in Space: From Science Fiction to the International Space Station (Caprara), 40
Lofgren, Gary, 12–13
Lower Torso Assembly (LTA), 60, 63
Lu, Ed, 50
lungs, 68

Marshall, Bob, 94
Marshall Space Flight Center, 10
mattress, 54
McCain, John, 97
McNamara, Robert, 13
"Medical Emergencies in Space" (Czarnik), 74
medicine. *See* space medicine
Microgravity Science Glovebox, 85, 88

micrometeoroid shield, 26–27
microsurgery, 86
military reconnaissance, 91, 93
Miller, Karen, 66
Mir
  building of, 31–36
  death of, 37
  health effects from living on, 72–73, 75
  research conducted on, 79, 81, 86, 90
Mir Space Station: A Precursor to Space Colonization, The (Harland), 35–36
Mission Control, 74
mistakes, learning from, 23–26
modules, 32–34, 40–44
Moon, 10, 12
motion sickness, 48–49
Muratore, John, 47
Musgrave, Mary E., 45, 89
music, 66
mutiny, 73, 75
Mylar, 60

Nadir window, 82
National Aeronautics and Space Administration (NASA)
  develops escape vehicle, 47
  develops shuttle, 30
  rejects Tito's offer to tour space, 92
  reveals budget for ISS, 96
  von Braun transferred to, 10
nightmares, 55
Node 1, 42, 44
Nomex, 60
Noordung, Herman, 9
nucleation, 87–88
nylon, 60

Oberth, Hermann, 9
Ochoa, Ellen, 66
optical power transmission, 86
oxygen, 12, 68

Park, Robert, 96
Pettit, Don, 62

photographs, 29, 79–80, 91
photovoltaic cells, 22
plant growth, 89–91
plasma crystals, 57
Pogue, Bill, 73
Popular Mechanics (magazine), 47
power, 21–23
pressure suits, 12, 67
Primary Life Support System (PLSS), 60, 63
Priroda (module), 81
Progress, 34–35, 37
Proton rocket, 32–33, 42
psychology, 73–76

radar, 13
radiation, 59, 61
red blood cells, 68
research
  astronomy, 78–80
  cellular, 83–84
  creating materials in weightless environments, 84–87
  environmental hot spots, 80–81, 83
  military reconnaissance, 91, 93
  new discoveries, 87–89
  value of, 94–97
Research Projects Laboratory, 14
resupply vehicles, 34–36, 41
right ventricle, 68
robotic arm, 45–46
rocket, 12, 17, 32–33, 35, 42
rocket team, 10
Roemer, Tim, 96
Romanenko, Yuri, 71–72, 74

Salyut
  assembly of, 17
  cardiovascular system changes on, 67–68
  death of, 37
  internal space in, 17
  launching of, 13, 15
  learning from mistakes on, 23–26
  research conducted on, 79

solar panels on, 22–23
space walks from, 74
successes with, 30–31
satellite, 11
Saturn launch vehicle, 10
sauce, 50
scatterometer, 80–81
Scott, Karen, 82
seed-to-seed cycling, 89
Service Module, 50
Shapiro, Jay, 71
shaving, 58
shoes, 20–21
shower, 58
shuttle, 30
Silber, Paul, 84
Skylab
  assembly of, 17
  bread crumbs on, 51
  exercise equipment on, 55–56
  health effects from living on, 67–68, 70
  internal space in, 17–18
  launching of, 15, 26–29
  research conducted on, 79–81, 90
  shower stall in, 58
  solar panels on, 17, 22–23
  space suits used on, 60
sleep, 53–55
Smith, Marcia, 93
smoke, 83
snoring, 55
sodium, 86
solar panels, 17, 22–23, 27–28, 41–42, 96
solar winds, 29
songs, 55
Soyuz, 23–25, 33, 46–47
space adaptation syndrome, 48–49
space medicine
  blood and fluid distribution, 65, 67
  cardiovascular system changes, 67–69
  muscular changes, 72–73

Index

music as relaxant, 66
psychological effects of space life, 73–76
skeletal system changes, 69–72
space race, 11–14
"Space Station Music" (Miller), 66
space stations
  architecture of, 16–17
  death of, 37
  for defense, 13–14
  first launch of, 13, 15
  internal space in, 17–18
  learning from mistakes on, 23–26
  power for, 21–23
  value of research from, 94–97
  weightlessness in, 18–21
  writers' visions of, 9
space suits, 60, 63
space tourism, 92
space walks, 61–64, 74
Special Purpose Dexterous Manipulator, 46
spoon, 50
Sputnik (satellite), 11
stationary bikes, 55
straps, 21
stress, 73, 75
Submillimetron, 79–80
Sun, 23, 27–29, 59, 61
sunspots, 29, 79

table, 50
teeth, brushing, 57–58
Teflon, 60
telescopes, 79–80
tether, 74
thermostabilized, 51, 53
Thompson, Robert F., 30–31
Tito, Dennis, 92
toilet, 58–59
tortillas, 50
tourists, 92

treadmills, 55

Unity Node, 42, 44
U.S. Army, 10
utensils, 50

vacuum, 12
Velcro, 19
ventricles, 68
Verne, Jules, 10
vomit, 49
von Braun, Wernher, 9–10
V-2 ballistic missile, 10

walls, 19–20
Walz, Carl, 66
Waring, Stephen, 14
water, 12, 35, 50–51, 57–58
weightlessness
  creating materials in environments of, 84–87
  described, 19
  effects of, on eating, 50
  effects of, on snoring, 55
  effects of, on the body
    blood and fluid distribution, 65, 67
    cardiovascular system, 67–69
    heart, 56
    muscular changes, 72–73
    skeletal system changes, 69–72
  engineering for, 18–21
  growing plants in, 89–91
  understanding of, 12
Wells, H.G., 10
White, Ed, 74
windows, 82
World War II, 10

X-38, 47

zinc oxide, 86
zirconium, 86

Picture Credits

Cover: © Stocktrek/CORBIS
AP/Wide World Photos, 92
© Bettmann/CORBIS, 8, 25, 59
EPA/Landov, 69
Hulton/Archive by Getty Images, 20, 28
Johnson Space Center/NASA, 14
Chris Jouan, 22, 26, 34, 38, 43, 53, 63, 71, 88
Marshall Space Flight Center/NASA, 10
NASA, 16, 31, 39, 44, 54, 66, 74, 75, 78, 82, 85, 91, 95

About the Author

James Barter received his undergraduate degree in history and classics at the University of California, Berkeley, followed by graduate studies in ancient history and archaeology at the University of Pennsylvania. Mr. Barter has taught history as well as Latin and Greek.

A Fulbright scholar at the American Academy in Rome, Mr. Barter worked on archaeological sites in and around the city as well as on sites in the Naples area. Mr. Barter also has worked and traveled extensively in Greece.